STRANGE &
UNEXPLAINED
HAPPENINGS
When Nature Breaks
the Rules of Science

STRANGE &
UNEXPLAINED
HAPPENINGS

When Nature Breaks
the Rules of Science

volume 2

Jerome Clark and Nancy Pear

AN IMPRINT OF GALE RESEARCH,
AN INTERNATIONAL THOMPSON PUBLISHING COMPANY.

Changing the Way the World Learns

NEW YORK • LONDON • BONN • BOSTON • DETROIT • MADRID
MELBOURNE • MEXICO CITY • PARIS • SINGAPORE • TOKYO
TORONTO • WASHINGTON • ALBANY NY • BELMONT CA • CINCINNATI OH

STRANGE AND UNEXPLAINED HAPPENINGS:
When Nature Breaks the Rules of Science

Jerome Clark and Nancy Pear, *Editors*

STAFF

Sonia Benson, *U•X•L Developmental Editor*
Kathleen L. Witman, *U•X•L Associate Developmental Editor*
Carol DeKane Nagel, *U•X•L Managing Editor*
Thomas L. Romig, *U•X•L Publisher*

Margaret A. Chamberlain, *Permissions Associate (Pictures)*
Shanna P. Heilveil, *Production Associate*
Evi Seoud, *Assistant Production Manager*
Mary Beth Trimper, *Production Director*

Mary Krzewinski, *Art Director*
Cynthia Baldwin, *Product Design Manager*
Terry Colon, *Illustrator*

♾️™ This book is printed on acid-free paper that meets the minimum requirements of American National Standard for Information Sciences—Permanence Paper for Printed Library Materials, ANSI Z39.48-1984.

ISBN 0-8103-9780-3 (Set)
    0-8103-9781-1 (Volume 1)
    0-8103-9782-X (Volume 2)
    0-8103-9889-3 (Volume 3)

Printed in the United States of America

I(T)P™

U•X•L is an imprint of Gale Research Inc., an International Thomson Publishing Company. ITP logo is a trademark under license.

10 9 8 7 6 5 4 3 2 1

Contents

VOLUME 1

I
UFOs: The Twentieth-Century Mystery 1

II
Ancient ETs and Their Calling Cards 23

VII
Light Shows 101

VIII
Strange Showers:
Everything but
Cats and Dogs 119

IX
More Weird Weather 147

XIII
Shaggy, Two-footed Creatures Abroad 263

XIV
Extinction Reconsidered 293

XV
Other Fantastic Creatures 327

VOLUME 3

XIX
Other Strange Events 487

READER'S GUIDE

Scope

Strange and Unexplained Happenings: When Nature Breaks the Rules of Science is a reliable reference guide to *physical* phenomena, as opposed to psychic or supernatural phenomena. It picks up where the studies of the occult and parapsychology leave off, telling the rest of the story of the world's mysteries—those dealing with strange natural and quasi-natural phenomena, things that seem to be a part of our world but are usually disputed or ignored by conventional science. No one who reads the newspaper or watches television is unfamiliar with reports of UFOs, the Loch Ness monsters, or Bigfoot. In *Strange* the stories of these and many more anomalies (abnormal or peculiar happenings) are presented with enough detail to stimulate wonder and surprise in even the most skeptical reader.

Research into such anomalies is not confined to a small group of nontraditional scientists. Academic teams, popular writers and members of anomalists' societies have examined strange physical phenomena in works ranging from sober, scientifically based analysis to wild conjecture. *Strange and Unexplained Happenings* presents the ideas of most of the major players in the field of each particular phenomenon, whether scientific or sensational. The entries also provide clear explanations of the kind of theory and research that have been applied. The greatest value of these fascinating accounts may not be in the answers they provide, but in the important questions they provoke.

Features

The three volumes of *Strange and Unexplained Happenings* are presented in chapters arranged by subject. Thus a student can look up Bigfoot and then browse the entire Shaggy, Two-footed Creatures in North America chapter, where he or she will find many other types of hairy bipeds that may or may not be related to the Bigfoot sightings. Similarly, by looking up "Flying Humanoids," the reader will find a variety of examples of visitors from other worlds.

Since anomalies, like most things in life, do not always fit neatly into one category or another, the volumes are extensively cross-referenced. Within each entry the names of related phenomena that have their own entries elsewhere in the set appear in boldface for quick reference. A thorough cumulative subject index concludes each volume.

The language used by anomalists (those who study strange happenings) presents a colorful variety of scientific and pseudoscientific terms. Although most of the terms are defined within the text, the glossary of terms appearing in the frontmatter of each volume ensures easy accessibility for all readers. Boxed material within the entry also provides interesting facts and explanations of concepts and terminology.

Strange centers on phenomena for which evidence is usually insufficient, if not altogether lacking. Photographs and drawings of all kinds have appeared as "proof" that some strange thing exists or happened. The photographs are often blurry and the drawings are often crude but, as the only evidence available, they are an important part of the history of the phenomenon and thus have been included in these volumes whenever possible.

Brief biographies of some of the foremost anomalists, whose works are most frequently cited within these volumes, are set off from the text in boxes. "Reel Life" boxes feature some modern and time-honored movies that have made such mysteries as werewolves, sea monsters, and UFOs a part of our culture.

Key sources are provided at the end of each entry. A Further Investigation section at the conclusion of each volume provides an annotated listing of the major books, periodicals, and organizations the student may wish to consult in further research into a particular strange phenomenon.

Comments and Suggestions

We welcome your comments on this work as well as your suggestions for strange events to be featured in future editions of *Strange and Unexplained Happenings*. Please write: Editors, *Strange and Unexplained Happenings,* U·X·L, 835 Penobscot Bldg., Detroit, Michigan 48226-4094; call toll-free: 1-800-877-4253; or fax: 313-961-6348.

INTRODUCTION

Strange and Unexplained Happenings: When Nature Breaks the Rules of Science is a book about *anomalies,* human experiences that go against common sense and break the rules that science uses to describe our world. In the words of folklorist Bill Ellis, "Weird stuff happens." *Strange and Unexplained Happenings* takes a look at the "weird stuff" that abounds in the reports of ordinary people who have had extraordinary experiences. Accounts of flying saucers, reptile men, werewolves, and abominable snowmen grab our attention and send shivers down our spines. When similar strange accounts are repeated time and again by different witnesses in different times and places, they capture the attention of the scientific community as well.

The three hardest words for human beings to utter are *I don't know.* Because we like our mysteries quickly and neatly explained, in modern times we have come to ask scientists to find logical explanations for strange human experiences. Sometimes science can use its knowledge and tools to find the answers to puzzling incidents; at other times it offers explanations that don't seem to fit the anomalies and only add to the confusion about them. When experiences are especially unbelievable, scientists may simply decide that they never really happened and refuse to consider them altogether. Most of us believe that as science learns more, it will be able to explain more. Still, it is almost certain that science will never be able to account for all the "weird stuff" that human beings encounter.

When an anomaly is reported, it is natural not to believe it, to be skeptical. One usually wonders about the witness. Could the person be lying for some reason? Tricks and hoaxes do occur. There are people who go to great lengths to fool scientists and the public, who hope to find fame and fortune by false claims or simply to prove to themselves how clever they are. Photographs of extraordinary happenings are often fake; it is thought that nearly 95 percent of all UFO photos are false, and some of the best film footage of the Loch Ness monster is judged doubtful as well. As a matter of fact, most investigators of anomalies feel that a lot of photographs of an incident signals a fake, because: (1) most people don't walk around with cameras ready to snap strange sights; (2) people having weird or scary experiences are often in shock or terror, and taking pictures is the last thing on their minds; and (3) anomalies generally last for just a matter of seconds. Investigators also believe that the fuzzier the photo, the more likely it is to be real, because pictures taken by people with shaking hands rarely turn out clearly, while hoaxers know that poor photographs won't get the results they are looking for.

It is also natural to wonder about the mental health of a person who has witnessed an extraordinary happening. Common sense tells us that *all* weird accounts should be blamed on the poor memories, bad dreams, or wild imaginings of confused and unwell minds! Still, psychologists who have examined witnesses of anomalies find them, for the most part, to be the same as people who have had no odd experiences at all. Also, the sheer number of strange reports rattles our common sense a bit, as do cases of multiple witnesses, when large groups of people observe the same strange happenings together.

More interesting still are accounts that have been repeated for centuries; reports of lake monsters in the deep waters of Loch Ness, for example, began way back in A.D. 565! Interesting, too, are reports that are widespread. The Pacific Northwest region of North America has its Bigfoot sightings, western Mongolians tell stories about the Almas, and accounts of the yeti have been reported in the high reaches of the Himalayas. While the languages and cultures surrounding these legends may differ, it is clear that witnesses are describing a similar creature: a hairy, two-legged "apeman." When observers report sightings of sea serpents they may describe them as smooth and snakelike or maned like horses, with many humps or finned like fish. Even when details vary widely, it is difficult to ignore their basic sameness: all suggest the existence of large, as yet unknown, sea-going animals.

It is true that in many cases of strange happenings, people have been misled or mistaken. The Bermuda Triangle, an area of the Caribbean where ships and planes were reported to mysteriously disappear, for example, was considered a real threat for more than two decades. That is, until weather records and other documents were properly researched, proving that the location was as safe as any other body of water. In the same way, the strange cattle "mutilations" that worried farmers in Minnesota and Kansas in the 1970s—and stirred up all sorts of weird explanations—required the special skills of veterinary pathologists to find that the cause was a simple, but gruesome, infection. Sometimes strange accounts do seem to change and grow as they are reported over the decades and in print. Human beings do want the truth ... but they also like a good story!

Science, too, has made its mistakes over the years. When sailors gave accounts of large sea creatures with giant eyes and many tentacles they were told that they were seeing floating trees with large roots. We now know that their accounts described giant squids. Gorillas and meteors were also rejected by scientists not that long ago!

But then, of course, some anomalies are more believable than others. When an odd happening turns the way we think about the world upside down it is described as "high-strange"; less weird accounts are lower on the strangeness scale. It is not *completely* unthinkable that unknown creatures still exist in some remote regions of the globe as many cryptozoologists (people who study "hidden" animals) believe. Wildlife experts and marine biologists may, over time, find that creatures like Bigfoot and "Nessie" are real. In the same way, physicists and meteorologists may find the reasons for ball lightning, or for the mysterious ice chunks that fall from the sky. The discovery of intelligent beings from outer space, on the other hand, would really shake us up and force us to rethink our lives and our place in the universe. As high on the strange scale as this idea is, though, there is enough hard evidence—like odd radar trackings and soil samples from UFO landing sights—to make it worth considering.

Strange accounts, no matter how farfetched, deserve some careful thought. Although most readers set their own limits as to how high on the strange scale they can go, the kinds of questions raised by anomalies are worth pursuing, even if the event or object is beyond one's own limits of belief. True understanding of anomalies takes time, effort, and an open—but not a gullible—mind. *Strange and Unexplained Happenings* doesn't deal with belief or disbelief. It only shows that human experiences come in more shapes and sizes than we could ever imagine!

GLOSSARY

A

anomaly: something that is abnormal and difficult to explain or classify by conventional systems. An *anomalist* is someone who studies or collects anomalies.

anthropoid: ape.

anthropology: the study of human beings in terms of their social relations with each other, their culture, where they live, where they come from, physical characteristics, and their relationship with the environment.

archaeology: the scientific study of prehistory by finding and examining the remains of past life, such as fossils, relics, artifacts, and monuments.

arkeology: a term used to describe the search for the remains of Noah's Ark at the site where it landed after the Great Flood, as chronicled in the Book of Genesis in the Bible.

astronomy: the study of things that are outside of the Earth's atmosphere.

Atlantis: a fabled island in the Atlantic inhabited by a highly advanced culture. According to Greek legend an earthquake caused the island to be swallowed up by the sea. Some still believe in the legendary Atlantis today, and societies have arisen in order to actively search for its remains.

atmospheric life forms: *See space animals.*

B

bioluminescent organisms: plants and animals that make their own light by changing chemical energy into light energy. Bioluminescent organisms are especially common in places where no light penetrates, like the depths of the ocean.

bipeds: animals that walk on two feet.

C

CE1: a UFO seen at less than 500 feet from the witness.

CE2: a UFO that physically affects its surroundings.

CE3: a being observed in connection with a UFO sighting.

cereology: the study of crop circles.

coelacanth: a large fish that, until 1938, had only been known through fossil records and was thought to have been extinct for some 60 million years. In 1938 a coelacanth was caught in the net of a South African fishing boat, giving rise to speculation that other species that had been officially declared extinct may live on.

contactee: a person who claims to have ongoing communications with one or more extraterrestrials. A *physical contactee* claims to have had actual physical contact with extraterrestrials and often will produce photographs or other material evidence of these meetings. A *psychic contactee* claims to have received messages from space in dreams or through automatic writing (writing performed without thinking, seemingly directed by an outside force).

corpse candles: also called death-candles; lights appearing in the form of a flame or a luminous mass, according to folk tradition, that foretell an impending death.

cover-up: an attempt made by an organization or group to conceal from the public the group's actions or information it has received or collected.

creationism: the belief, based on a word-for-word reading of the Bible's Book of Genesis, that God created all matter, all living things, and the world itself, all at the same time and from nothing.

creation myths: sacred stories that explain how the Earth and its beings were created.

Cro-Magnon race: a race that lived 35,000 years ago and is of the same species as modern human beings (*Homo sapiens*). Cro-Magnons stood straight and were six or more feet tall; their foreheads were high and their brains large. Skillfully made Cro-Magnon tools, jewelry, and cave wall paintings suggest that the Cro-Magnon race had an advanced culture.

cryptozoology: the study of lore concerning animals that science does not account for, including animals thought to be extinct, animals that have been seen only by local populations, or animals thought to exist only in certain areas that show up elsewhere. The objective of cryptozoology is generally to evaluate the possibility of these animals' existence.

D

debunk: to expose something as false or as a hoax.

dowsing: a folk method for finding underground water or minerals with a divining rod. The divining rod is usually a forked twig; the "diviner" holds the forked ends close to his or her body, and the stem supposedly points downward when he or she walks over the hidden water or desired mineral. Some believe that dowsing can be used as a method to predict when and where a crop circle will appear.

E

ethereans: fourth dimensional human beings; a theoretical group of beings like humans, only more advanced, who live in another (or fourth) dimension that coexists with our world. Just as the stars and planets of our universe have their etheric counterparts, ethereans are human beings in a different reality.

evolution: a process in which a group of plants or animals—such as a species—changes over a long period of time, so that descendants differ from their ancestors. Theoretically the changes result from *natural selection,* a process in which the strongest and the most adept at survival pass on their characteristics to the next generations. Characteristics that make group members less successful at surviving and breeding are slowly weeded out.

extinct: no longer in existence.

extraterrestrial: something that came into being or lives outside of the Earth's atmosphere, or something that happened outside of the Earth's atmosphere.

F

Fortean: an adjective used to describe outlandish, sometimes sarcastic, and generally antiscientific theories in the literature of the strange and unexplained. The word is derived from the pioneer of physical anomalies, Charles Fort, who frequently poked fun at the weak attempts science made to explain away strange events by offering wacky theories of his own.

G

geophysics: a branch of earth science dealing with physical processes and phenomena occurring within or on the Earth.

H

hallucination: an illusion of seeing, hearing, or in some way becoming aware of something that apparently does not exist in reality.

herpetology: the scientific study of reptiles and amphibians.

humanoid: having human characteristics; a being that resembles a human.

I

ichthyology: the study of fish.

inorganic: composed of matter other than plant or animal; relating to mineral matter as opposed to the substance of things that are or were alive.

L

Lemuria: a legendary lost continent in the Pacific Ocean somewhere between southern Africa and southern India. Unlike accounts of Atlantis, which date back to the writings of Plato in ancient Greece, theories about Lemuria arose in the nineteenth century in the doctrine of occultists such as Madame Helena Petrovna Blavatsky, cofounder of the Theosophical Society, Max Heindel, founder of the Rosicrucian Fellowship, Rudolf Steiner, founder of the Anthroposophical Society, and Theosophist W. Scott-Elliott.

lycanthropy: the transformation of a man or woman into a wolf or wolflike human.

M

meteor: one of the small pieces of matter in the solar system that can be seen only when it falls into the Earth's atmosphere, where friction may cause it to burn or glow. When this happens it is sometimes called a "falling" or "shooting" star.

meteorite: a meteor that survives the fall to Earth.

meteoroid: any piece of matter—ranging in mass from a speck of dust to thousands of tons—that travels through space; it is composed largely of stone or iron or a mixture of the two. When a meteoroid enters the Earth's atmosphere it becomes visible and is called a *meteor.*

meteorology: the science of weather and other atmospheric phenomena.

mollusks: the second largest group of invertebrate animals (those without a backbone). They are soft-bodied, and most have a distinct shell. Mollusks usually live in water and include scallops, clams, oysters, mussels, snails, squids, and octopuses.

mutology: the investigation of cattle mutilations.

N

Neanderthal race: a species that lived between 40,000 and 100,000 years ago. Neanderthal remains have been found in Europe, northern Africa, the Middle East, and Siberia. The classic Neanderthal man had a large thick skull with heavy brow ridges, a sloping forehead, and a chinless jaw. He was slightly over five feet tall and had a stocky body. The link between Neanderthal man and modern human beings is unclear. Many anthropologists believe that they evolved separately from an earlier common ancestor.

New Age: a late-twentieth-century social movement that draws from American Indian and Eastern traditions and espouses spirituality, holism, concern for the environment, and metaphysics.

O

occultism: belief in or study of supernatural powers.

OINTS (Other Intelligences): a term coined by biologist Ivan T. Sanderson that includes not only extraterrestrials, but also space animals, undersea civilizations, poltergeists, and extradimensional beings.

organic: composed of living plant or animal matter.

ornithology: the scientific study of birds.

P

paleontology: the scientific study of the past through fossils and ancient forms of life.

paracryptozoology: the study of animals whose existence, even to the most open-minded, seems impossible (*paracryptozoology* means "beyond cryptozoology").

paranormal: not scientifically explainable; supernatural.

paraphysical: a combination of the terms *paranormal* (outside the normal) and *physical*. This concept, used by some anomalists, encompasses both natural occurrences (like leaving tracks) and unnatural occurrences (like disappearing instantly).

plesiosaurs: a suborder of prehistoric reptiles that dominated the seas during the Cretaceous Period (136 to 65 million years ago). Their bodies were short, broad, and flat. They had short pointed tails. Their small heads were supported by long slender necks, ideal for darting into the water to catch fish. Plesiosaurs swam with a rowing movement, using their four powerful, diamond-shaped flippers like paddles. They were often quite large, measuring up to 40 feet in length.

porphyria: a rare genetic disease often linked with werewolf sightings. Porphyria sufferers are plagued by tissue destruction in the face and fingers, open sores, and extreme sensitivity to light. Their facial skin may take on a brownish cast, and they may also suffer from mental illness. The inability to tolerate light, plus shame stemming from physical deformities, may lead the afflicted to venture out only at night. Some in the medical community have suggested that sightings of werewolves have really been of individuals with porphyria.

primates: a member of the group of mammals that includes man, apes, monkeys, and prosimians, or lower primates. Primates have highly developed brains and hands with opposable thumbs that are very adept at holding and grasping things.

psychosocial hypothesis: a belief that UFOs and other anomalies are powerful hallucinations shaped by the witness's psyche and culture, and that strange sightings are actually insights into deep realms of the human imagination rather than evidence of visitors from other worlds.

Q

quadrupeds: animals that walk on four feet.

R

radar: a method of detecting distant objects and determining their position, speed, or other characteristics by analyzing radio waves reflected from them.

S

saucer nests: circular indentations that one could imagine to be left by hovering or grounded spacecraft; saucer nests, found in the 1960s and 1970s before the current crop circle mania began, have many features comparable to crop circles, but the connection between the two phenomena has not been established.

sauropods: huge plant-eating reptiles with long necks and tails, small heads, bulky bodies, and stumplike legs; *Diplodocus, Apatosaurus* (*Brontosaurus*), and *Brachiosaurus* were sauropods.

shape-shifter (or shape-changer): one who can change form at will; in medieval and later chronicles, shape-shifting was associated with witchcraft, and such shape-shifters as black dogs and were-wolves were often considered to be either agents of the devil or Satan himself.

sonar: a method of tracking that uses reflected sound waves to detect and locate underwater objects.

space animals (atmospheric life forms): hypothetical life forms existing in the upper atmosphere. Several ufologists have suggested that UFOs are neither spacecraft nor cases of mistaken identity but *space animals.*

spontaneous generation: a once widespread belief that living things can spring from nonliving material; thus, when rain hits the ground it can give rise—out of the mud, slime, and dust—to all sorts of living matter.

supernatural: of or relating to an order of existence outside the natural, observable universe.

T

teleportation: the act of moving an object or person from one place to another by using the mind, without using physical means.

theosophy: teachings about God and the world based on mystical insight; in 1875 the Theosophy movement arose in the United States, following Eastern theories of evolution and reincarnation.

transient lunar phenomena (TLP): unusual, short-lasting appearances on the moon's surface typically observed by astronomers through telescopes, and more rarely by the naked eye.

U

UFO (unidentified flying object): a term first coined by a U.S. Air Force worker that came into common usage in the mid-1950s to describe the "flying saucers" or mysterious discs that were being observed in the air and were suspected by some to be the craft of extraterrestrial visitors.

UFO-abduction reports: the accounts of a significant group of witnesses who claim to have been kidnapped by aliens. Many witnesses described large-headed, gray-skinned humanoids who subjected them to medical examinations. Some witnesses experienced amnesia after their encounters and recalled them only through hypnosis.

ufology: the study of unidentified flying objects.

ultraterrestrials: beings from another reality.

W

water horse: a folkloric creature believed by many to be a dangerous shape-changer that can appear either as a shaggy man who leaps out of the dark onto the back of a lone traveler or as a young horse that, after tricking an unknowing soul onto its back, plunges to the bottom of the nearest lake, killing its rider.

waterspouts: funnel- or tube-shaped columns of rotating, cloud-filled wind, usually extending from a cloud down to the spray it tears up from the surface of an ocean or lake.

Z

zeuglodon: a primitive, snakelike whale thought to have become extinct long ago.

PICTURE CREDITS

STRANGE &
UNEXPLAINED
HAPPENINGS

When Nature Breaks
the Rules of Science

Cryptozoology

- THE SCIENCE OF "UNEXPECTED" ANIMALS

Cryptozoology

THE SCIENCE OF "UNEXPECTED" ANIMALS

The two best-known areas of investigation in unusual phenomena are ufology (the study of UFOs) and cryptozoology. The term *cryptozoology* comes from the Greek words *kryptos* (hidden), *zoon* (animal), and *logos* (discussion)—in short, it means the study of hidden animals. Bernard Heuvelmans, the man who named the field, favored the word "hidden" instead of "unknown" to describe these mystery animals. This is because he felt that such animals were usually known to the local people who lived near them, who often supplied the eyewitness accounts and information about appearance and behavior that would make its way into the larger world and capture the interest of scientists. Heuvelmans thought it would be better to call them animals "undescribed by science." Two of cryptozoology's most famous subjects are **Bigfoot** and the **Loch Ness monster.**

In 1982, at the Smithsonian Institution in Washington, D.C., the International Society of Cryptozoology (ISC) was created so that biologists interested in unknown, hidden, or doubted animals would have a formal organization through which research could be done. There the definition of cryptozoology was made even more clear. It was agreed that cryptozoology should include the possible existence of *known* animals in areas where they were not supposed to be, as well as the possible existence of animals thought to be extinct. "What makes an animal of interest to cryptozoology," the *ISC Newsletter* decided, "is that it is unexpected."

PTERODACTYLS.

PLATE VI.

LONG-NECKED SEA-LIZARD.

Plesiosaurus dolichodeirus. Length 22 feet.

CUTTLE-FISH.

Some cryptozoologists have found evidence of the modern-day existence of animals thought to have been extinct for millions of years, like the pterosaur and sea lizard (dinosaur) pictured here.

The Prehistory of Cryptozoology

In the eighteenth century Swede Carolus Linnaeus was the first to put plants and animals into modern, scientific classifications. He believed that God created the forms and functions of these plants and animals and that they never changed. (Charles Darwin would present another theory called evolution—based on gradual transformation of a species—more than a century later.) Animals that did not fit into the classifications put forth by Linnaeus, like unicorns or **sea serpents**, were no longer considered worthy of scientific study.

By the nineteenth century, scientists felt that they had found and classified nearly all the creatures on earth. In 1812 Baron Georges Cuvier, the great French naturalist remembered today as the father of paleontology (the study of the past through fossils), declared that all species of large quadrupeds—animals that walked on four feet—had been discovered. It was a belief shared by most other scientists. Nonetheless, thousands of new species of animals, including large quadrupeds, have been discovered since Cuvier's time.

The fierce-looking, humanlike *pongo,* known to us as the gorilla, was officially recognized in 1847. The **giant squid** (called the Kraken in folklore) and the giant panda were also discovered around that time. Because odd new creatures continued to appear, some scientists began to give the strange animals of legends and folklore a second look. The question of sea serpents, for example, captured the attention of scientists and writers who thought that the witnesses were too reliable and the sightings too detailed to be ignored. These creatures became a popular topic of discussion in European and American scientific journals in the nineteenth century.

Zoologist editor Edward Newman, for one, hoped for a shift in scientific thinking—a greater open-mindedness. Of the sea serpent question, he wrote in 1847, "A natural phenomenon of some kind has been witnessed; let us seek a satisfactory solution rather than terminate enquiry by the shafts of ridicule." Defenders of the sea serpent's existence included some important natural scientists. In 1892 Dutch zoologist Antoon Cornelis Oudemans collected all available reports and carefully examined them in a widely read book, *The Great Sea-Serpent.* Heuvelmans considered the volume the "true starting point of the new discipline" of cryptozoology.

But at the same time, powerful scientists—whose doubts would win out in the end—attacked sea serpent reports as lies and mistaken identifications (as many of them were) and poked fun at witnesses. And scientists who dared to disagree sometimes had to pay a high price. The highly regarded French-American zoologist Constantin Samuel Rafinesque, for example, found his reputation and career in ruins following his writings on the sea serpent issue.

In the twentieth century, there is a growing feeling that the discovery of large, still unknown animals is *very* unlikely. Yet such animals are still being found. The three major discoveries of our time—though far from the *only* discoveries—were of the okapi (1900), the mountain gorilla (1903), and the coelacanth (1938). The okapi, a short-necked African animal related to the giraffe but resembling more of a donkey-zebra mix, would eventually become the symbol of cryptozoology, appearing on the ISC's logo. The coelacanth, found in the net of a South African fishing boat, was believed to have become extinct some 60 million years ago. (Thought to be 400 million years old, the species was around long before the dinosaurs!) The discovery of the coelacanth is considered one of the twentieth century's most important zoological finds.

That discovery made some scientists and writers wonder: if the prehistoric coelacanth could be found, what other strange animals had

The coelacanth, found in the net of a South African fishing boat, was believed to have become extinct some 60 million years ago.

escaped extinction? A few, who had heard reports from Africa of a strange sauropod-like animal (like the brontosaurus) called the **mokele-mbembe,** suggested that perhaps dinosaurs were not extinct after all. Other reports from the continent sounded like sightings of another prehistoric reptile, the flying **pterosaur**. In the January 3, 1948, issue of the popular magazine the *Saturday Evening Post,* biologist Ivan T. Sanderson brought these questions before the public in his article "There Could Be Dinosaurs." Willy Ley, educated in paleontology and zoology, also wrote about unknown animals in magazine articles and books for the general public. But worldwide interest in strange animals had already been stirred in 1933, when the first photo of the Loch Ness monster appeared. Hundreds of magazine and newspaper articles and books were written about the creature, either making the case for or against its existence.

Heuvelmans and Modern Times

The most important cryptozoological book of all time, *On the Track of Unknown Animals,* appeared in French in 1955 and in English in 1958. Its author, Franco-Belgian zoologist Bernard Heuvelmans, had spent years collecting reports of strange animals from a wide variety of popular and scientific literature. *Unknown Animals,* a thick book that addressed only land-based creatures, did not use the word "cryptozoology." In fact, Heuvelmans did not coin the phrase until after his vol-

Bernard Heuvelmans

(1916-)

Two books sparked French cryptozoologist Bernard Heuvelmans's interest in strange animals during his youth: Jules Verne's *Twenty Thousand Leagues Under the Sea,* with its momentous giant squid battle scene, and Sir Arthur Conan Doyle's *The Lost World,* with its living dinosaurs. Later, as a successful science writer and zoologist, Heuvelmans read Ivan T. Sanderson's (also see entry: Hairy Bipeds in North America) 1948 article "There Could Be Dinosaurs" and decided then and there to focus his studies on the mysterious animals that so fascinated him.

His first book, *On the Track of Unknown Animals* (1955), brought together all the printed sources on unexpected land animals that Heuvelmans could find. The book, though scientific in its approach, was written to captivate the largest possible audience. It sold about one million copies worldwide. In 1968 Heuvelmans published *In the Wake of the Sea-Serpents,* an in-depth examination of giant ocean creature sightings. His numerous other books have been published only in French.

Heuvelmans coined the term *cryptozoology* and he is considered the most influential writer in this field. He established the Center for Cryptozoology in southern France in 1975. In 1982 he participated in the founding of the International Society of Cryptozoology (ISC) and serves as its president from his home in France.

ume was published. But he began using it when writing letters, and in 1959 a friend dedicated a book to Heuvelmans, the "master of cryptozoology." Thus the word appeared in print for the first time.

Translated into many languages, *Unknown Animals* sold a million copies worldwide and almost single-handedly turned the study of such beasts into a science all its own. Other events led to interest in the subject as well. In the 1950s, magazines with huge readerships like *Life* and *National Geographic* ran stories on the **Yeti** (a Himalayan Bigfoot also known as the "abominable snowman"). And later in the decade, sightings and tracks of Bigfoot in America's Pacific Northwest attracted much attention.

The 1960s unleashed a flood of cryptozoological writings that continues to this day. And the manner in which writers approached these zoological mysteries varied widely—from careful and detailed scientific studies to the most breathlessly outlandish tales and theories. Sometimes this antiscientific point of view was described as Fortean, after **Charles Fort** (also see entry: **Falls from the Sky**), a pioneer in the field of anomalistics, or unusual events, who sometimes poked fun at science's weak attempts to explain away strange events—and who offered wacky theories of his own.

But for the most part, the new era focused on field research and evidence. Professionals and amateurs alike prowled forests, jungles, beaches, rivers, and lakes looking for specimens or other clues—mostly with poor results. The newly founded International Society of Cryptozoology played an important role in such field investigations. When anthropologist Roy Wagner reported that residents of New Guinea were seeing creatures they described as half-human and half-fish, for instance, the ISC sent investigators to accompany Wagner. Eventually they determined that the dugong—a plant-eating water mammal also known as the sea cow—was the creature the natives were actually seeing (also see entry: **Ri**). Over the next ten years ISC members also traveled to the Congo in search of mokele-mbembe and to China looking for "wild men," that country's version of Bigfoot. Other field research was done at Lake Champlain, the supposed home of water monster **Champ**, and in the Oregon/Washington wilderness where Bigfoot reportedly roamed.

The ISC also looked at less spectacular but still interesting "cryptids" (as puzzling animals are called), like the **thylacine**, an Australian flesh-eating marsupial said to be extinct but still reported in sightings, and the Eastern cougar. ISC secretary J. Richard Greenwell even obtained, from the mountains of western Mexico, a slain specimen of a large cat identified as the **onza**, which was mentioned in legends but the existence of which had long been doubted. Greenwell's fantastic cryptozoological find, however, became a disappointment when cursory laboratory studies produced no evidence that the creature was genetically different from other big cats native to the area.

As the ISC became more active, cryptozoology gained the support of some important scientists—and drew the attacks of others. Most of its critics believed, like Cuvier back in 1812, that no significant large animals were left to be discovered. In the late 1980s, while speaking at a zoological meeting, Greenwell found himself openly mocked! But that was unusual, for most scientists are aware of the field of cryptozoology and seem willing to hear it out. *Cryptozoology,* the ISC's yearly journal, publishes the work of well-respected biologists. Even contributors who are doubtful that cryptozoology's efforts will bring any real results admit that—despite the long odds—the search is still worthwhile. Of course the discovery of a specimen, living or dead, of a spectacular "unexpected animal" would make cryptozoology acceptable at once.

Paracryptozoology

Paracryptozoology addresses cases of unexpected animals the existence of which, even to the most open-minded, seems *totally* impossible ("paracryptozoology" means "beyond cryptozoology"). Nearly all the cryptids considered by Heuvelmans, the ISC, and other cryptozoologists trained in the biological sciences are thought to live in remote or thinly populated regions or in deep bodies of water. In places like these the existence of large unknown animals is at least a slim possibility.

But of course, these are not the only places where such creatures are sighted. Bigfoot and other similar-looking **hairy bipeds** (animals that walk on two feet) have been reported all over the United States and Canada, not just in the Pacific Northwest wilderness. And "monsters" of other sorts—from **werewolves** to **reptile men**—have been frequently reported in newspapers and other writings. Though witnesses who see these very strange creatures seem as solid and as normal as those who report more ordinary "cryptids," cryptozoologists have stayed clear of them and their claims. Most science-based cryptozoologists ignore paracryptozoology because it asks them to stretch reality a little *too* far.

Sources:

Bauer, Henry H., *The Enigma of Loch Ness: Making Sense of a Mystery,* Urbana, Illinois: University of Illinois Press, 1986.

Bord, Janet, and Colin Bord, *Alien Animals,* Harrisburg, Pennsylvania: Stackpole Books, 1981.

Clark, Jerome, and Loren Coleman, *Creatures of the Outer Edge,* New York: Warner Books, 1978.

Heuvelmans, Bernard, *In the Wake of Sea-Serpents,* New York: Hill and Wang, 1968.

Heuvelmans, Bernard, *On the Track of Unknown Animals,* New York: Hill and Wang, 1958.

Holiday, F. W., *The Dragon and the Disc: An Investigation into the Totally Fantastic,* New York: W. W. Norton and Company, 1973.

Keel, John A., *Strange Creatures from Time and Space,* Greenwich, Connecticut: Fawcett Publications, 1970.

Misplaced Animals

- WANDERING KANGAROOS

- BLACK PANTHERS AND OTHER STRANGE CATS

- BEAST OF EXMOOR

- BEAST OF GEVAUDAN

- ALLIGATORS IN SEWERS

- ENTOMBED ANIMALS

- NORTH AMERICAN APES

Misplaced Animals

WANDERING KANGAROOS

Kangaroos live only in the southern part of the world—in Australia, New Guinea, Tasmania, and surrounding islands. Yet oddly enough, the animals have been sighted living wild in the United States for nearly a century! While no one knows exactly how they got here, the number of sightings—and the witnesses reporting them (many being police officers)—make it almost certain that a few kangaroos have made America their home.

The reports began in 1899, when a Richmond, Wisconsin, woman saw a kangaroo run through her neighbor's yard. Because a circus was in town, she assumed that the animal had escaped. But, in fact, the circus had no kangaroo! The following year, in Mays Landing, New Jersey, a farm family heard a scream coming from near their barn. From what they could tell, the thing doing the screaming was a 150-pound kangaroo! From then on the family often saw the animal's tracks, eight to ten feet apart, leading to a large cedar swamp at the rear of their property.

In 1934 a "killer kangaroo" terrified Tennessee countryfolk, attacking and killing dogs, geese, and ducks. Because kangaroos are rarely violent—and never eat meat—the accounts were reported in newspapers around the country poking fun at area residents. Still, the witnesses stood by their stories and the local newspaper (the *Chattanooga Daily Times*) defended them, stating, "There is absolutely no doubt about these facts, a kangaroolike beast visited the community and killed dogs right and left, and that's all there is to it." However, because no other "killer" kangaroos have since been reported, this Tennessee case is still in question.

Two men saw and photographed this huge kangaroo in April 1978 as it lurked in the bushes along a Wisconsin highway.

There have been other kangaroo sightings. The headlights of a Greyhound bus driven by Louis Staub shone on a kangaroolike animal as it leaped across a highway near Grove City, Ohio, in 1949. Residents of Coon Rapids, outside Minneapolis, reported numerous sightings of kangaroos, sometimes traveling in pairs, from 1957 through 1967. And in 1958, kangaroos were sighted around several Nebraska towns, some 100 miles apart. One witness, beer brewer Charles Wetzel, took advantage of the accounts, naming one of his products "Wetzel Kangaroo Beer."

Arresting a Kangaroo

Residents of cities surrounding and including Chicago experienced a number of kangaroo incidents in the autumn of 1974. They began on October 18 when a Chicago man called police at 3:30 A.M. about a kangaroo on his porch.

The two unbelieving officers were amazed to find the report true; not sure of what to do, officer Michael Byrne tried to handcuff the animal, which—according to the policeman—"started to scream and got vicious." A scuffle followed, with the five-foot-tall kangaroo kicking officer Leonard Ciagi hard and often! The policemen called for backup, but by then the kangaroo was escaping down the street at about 20

A scuffle followed, with the five-foot-tall kangaroo kicking officer Leonard Ciagi hard and often!

miles per hour. Over the next two or three weeks, more sightings were reported in cities in Illinois and Indiana, some at the very same time; a kangaroo also made a second appearance in Chicago. Finally, on November 25, farmer Donald Johnson of Sheridan, Indiana, made the last reported sighting of the year. Johnson spotted a kangaroo running down the middle of a deserted country road; when the animal spotted him, it jumped over a barbed-wire fence and disappeared into a field.

Kangaroos continue to be seen around the United States and in parts of Canada. Because none of the animals have ever been caught or killed, the experience of two Menomonee Falls, Wisconsin, men takes on special importance. On April 24, 1978, the pair snapped two Polaroid photographs of a huge kangaroo in the bush along a local highway. Loren Coleman, the leading expert on North American kangaroo sightings, felt the animal looked very much like a "Bennett's wallaby or brush Kangaroo, a native of Tasmania."

Sources:

Coleman, Loren, *Mysterious America,* Boston: Faber and Faber, 1983.
Shoemaker, Michael, "Killer Kangaroo," *Fate* 38,9, September 1985, pp. 60-61.

BLACK PANTHERS AND OTHER STRANGE CATS

Early one November morning in 1945, Wanda Dillard was driving along Highway 90 in southern Louisiana (then a narrow two-lane road) when she noticed a set of glowing red eyes at the edge of the woods ahead. Fearing that the animal might sprint into the middle of the road, she slowed down. She recalled what happened next:

> As I braked, this huge black cat left the woods, and streaked across in front of the headlights and down the embankment on the left. It was such a beautiful thing that as soon as I could maneuver a turn, I went back to see if I could get another look at him, although I expected him to be at least over in the next parish [county] by then. But lo and behold, there he was, red eyes and all, crouched at the edge of the woods just as if he were waiting to play tag with the next motorist.

A black panther.

Dillard related that as the cat got used to the headlights of her car, it "simply sat down like any common old house cat and began to groom himself." She watched the animal for some ten minutes before traffic forced her to move on. She described the animal as "jet black, very sleek looking and well muscled, with a long black tail that he wrapped around himself as he sat there." She added that while "a bit smaller," it resembled a "mountain lion in both looks and movement."

Bill Chambers, a farmer in Champaign County, Illinois, came upon a similar sight nearly two decades later. While driving along just before sunset, he spotted a large cat in a nearby field. Moving to within 190 yards of it, Chambers watched through his binoculars, hoping to kill it with a clear shot. He would have no luck, however, because the cat was almost hidden by tall clover. Still, Chambers could clearly see its tail and noted that it appeared "jet black except for two tawny streaks under the jaws." The next day the farmer returned to the site and from the size and height of the clover patch guessed that the cat was about 14 inches tall at the shoulder and roughly four and a half feet long. Chambers also found tracks in the soft, wet ground. Nearly three inches across, "they had no claw marks like a dog would make."

The above sightings are extraordinary for several reasons. For one thing, during the last half of the nineteenth century, *Felis concolor*—a family of large cats commonly known as panthers, mountain lions, cougar, or puma—was hunted to extinction everywhere (except for a small population in Florida's Everglades) east of the Rocky Mountains. Secondly, these animals are almost *never* black!

Plenty of Panthers

Constant sightings—as well as some claimed (but unproven) killings—have led a number of wildlife specialists to conclude that a small population of *Felis concolor* has survived in the eastern United States and Canada. If nothing else, the argument about whether people are seeing big cats or mistaking large dogs or wild house cats for them has largely been concluded; few doubt that big cats are out there. But the question remains: Where did they come from? One theory suggests that pet owners bought panthers as kittens and secretly released them into the wild, where they grew up and out of control.

Of course, some investigators reject black panther reports. Robert L. Downing, who undertook a study of the subject for the U.S. Fish and Wildlife Service, trumpeted the traditional view: "Some black animals, such as Labrador retrievers, are reported to be black panthers because that's the color that panthers are supposed to be, according to folklore." His caution makes some sense—tracks reportedly belonging to black panthers *have* actually been proven to be prints of dogs.

But biologist Bruce S. Wright, director of the Northeastern Wildlife Station at the University of New Brunswick, Canada, became convinced that some of these black panther sightings deserved a closer look. Over the course of his research, he collected 20 reports that he felt were reliable. All occurred in daylight and at close range.

One of these cases is especially fantastic. It took place in Queens County, New Brunswick, on November 22, 1951. A man walking near his home around 6 P.M. "heard five loud yells off in the woods." After walking further he heard more yells and, turning around, saw a large animal come leaping at him. Unable to outrun the beast, the witness "had to stop and face it." He reported, "When I stopped, it stopped and stood up on its rear legs with mouth open and 'sizzling' and with forepaws waving." The witness swung the axe he was carrying at the animal but missed and began to run again. Meanwhile, the creature apparently returned to the woods. The witness described the animal as "black or dark grey in color. The tail was at least two and one-half feet long, and the animal was at least six feet long."

> "When I stopped, it stopped and stood up on its rear legs with mouth open and 'sizzling' and with forepaws waving."

Few doubt that big cats are out there. But the question remains: Where did they come from?

This account is remarkable for three reasons. One, the panther was black; two, it stood upright; and three, it was not afraid to attack a human being. "Real" panthers have learned through bitter experience to keep as far away as possible from their mortal enemy—humans—who nearly destroyed their species. They have been rendered fearful of people.

As unbelievable as the New Brunswick story might seem, a report half a continent away and nearly two decades later also included these three amazing details. It took place a mile south of Olive Branch, Illinois, on a dark, mostly deserted road that runs along the edge of the vast Shawnee National Forest. Mike Busby of Cairo, Illinois, was on his way to pick up his wife when his car stalled. As he was releasing the hood latch, he heard something off to his left. When he turned to look, he was startled to see two quarter-sized, almond-shaped, greenish, glowing eyes staring at him.

Suddenly the strange form—six feet tall, black, and upright—hit him in the face with two padded front paws. Busby fell, the animal on top of him, and as the two rolled around, it ripped his shirt to pieces and used its dull, two-inch claws to cut his left arm, chest, and stomach. He managed to keep its open mouth, with its long, yellow teeth, away from his throat. Though he never got a good look at the creature, Busby later said that he felt what seemed like whiskers around the mouth. Its deep, soft growls were like nothing he had ever heard before. In the lights of a passing diesel truck, the creature appeared to be a slick, shiny black and, for the first time, Busby saw the "shadow of a tail." The light seemed to frighten the animal, which loped across the road with "heavy footfalls" and disappeared into the woods.

Confused and in pain, the young man crawled back to his car. To his immense relief, it started without trouble. Busby drove to Olive Branch, where he met truck driver John Hartsworth, who said that he had seen Busby struggling with what looked like a *big* cat" but was unable to brake his vehicle and help. Later that evening Busby went to St. Mary's Hospital in Cairo to get a tetanus shot and relief for his pain. For days afterward he suffered dizzy spells and had trouble walking.

But black panther sightings number in the hundreds, perhaps even in the thousands. And they have occurred in places where neither *Felis concolor* nor any other big cat has *ever* existed. In areas where *Felis concolor* is known to exist—northern California or the southeastern United States—a large number of reported sightings were of black panthers. As a matter of fact, of the 615 big cat sightings collected between 1983 and 1990 by Eastern Puma Network News, some 37 percent of them were of black panthers. J. Richard Greenwell of the International Society of Cryptozoology found the situation "mind-boggling"; for despite all the reports, no one had ever come forward with a clear photo of a black panther, or a single skin.

Big Cats Sighted in Great Britain

There is one wildcat native to the British Isles—*Felis silvestris grampia.* A small cat that once roamed much of Great Britain, it now

Drawing of a creature spotted at Van Etten Swamp, New York, in the mid-1970s.

lives only in Scotland and possibly in some isolated regions of northern England. Still, Britain has had its share of big cat sightings, and it is very unlikely that *Felis silvestris* is responsible. Arguments and theories about the presence of big cats in Britain became particularly heated in the early 1960s, when the "Surrey puma" began to prowl the country lanes of southern England.

In the late summer of 1962, something described as "a young lion cub—definitely not a fox or a dog" was seen near a reservoir in a Hampshire park. Other sightings occurred but garnered no more than local attention. But that changed early on the morning of July 18, 1963, when a truck driver passing through Oxleas Wood, Shooters Hill, London, was startled to see a "leopard" leap across the road and into trees on the other side. Later that day four police officers had an even closer look: a "large golden animal" jumped over the hood of their squad car before disappearing into the woods. The only evidence turned up by a huge search involving dozens of officers, soldiers, and dogs were large footprints of a catlike animal.

Other reports from southeastern England followed. The cats were described as large panthers or pumas, either "fawn gold" or "black" in color. Deer, sheep, and cattle in the area were found slaughtered, with huge claw marks on their sides. One woman even claimed that a "puma" had struck her in the face with its two front paws as she was walking through a wooded area in Hampshire.

To wildlife biologists, all of this seemed impossible. Weighing early reports, Maurice Burton indicated the many reasons that made the whole episode so unbelievable. He noted that "from September 1964 to August 1966, the official records show 362 sightings," with perhaps just as many "claimed but not officially reported. In other words this animal [*Felis concolor*], declared by American experts to be 'rarely seen by man,' was showing itself on average once a day for a period of two years." Burton also noted that "in two years it was reported from places as far apart as Cornwall and Norfolk, over an area of southern England of approximately 10,000 square miles. It has even been in two places many miles apart at the same time on the same day." Another wildlife expert, Victor Head, remarked that a single panther would need to eat 250 British roe deer a year to survive; yet England's deer population showed no signs of such a loss.

Still, in the years to come, big cats—along with footprints and slain farm animals—would continue to be reported all over Britain. These accounts would try the patience of those who viewed the sightings as mistaken identifications, wild imagination, or trickery. These doubters would focus on the few recorded cases that *did* have logical explanations. For example, in August 1983 a Buckinghamshire woman saw a large pumalike black cat "wearing a studded collar"—suggesting, of course, that someone had once owned it before releasing it into the wild. In August 1975, in fact, a Manchester man had captured a leopard cub with a collar. And in Inverness, Scotland, a female puma was trapped in 1980. After a short time examiners concluded that it had been a pet for most of its life.

Yet it is hard to believe that Britain is crawling with big cats that are escaped pets. To be sure, experience shows that such creatures are

Later that day four police officers had an even closer look: a "large golden animal" jumped over the hood of their squad car before disappearing into the woods.

A "Surrey puma," photographed at Worplesdon, Surrey, from a distance of 35 yards in August 1966. The two ex-police photographers were sure it was not a wild tomcat.

poorly suited to the wild and usually starve to death or are recaptured or killed shortly after their release or escape. And if escaped pets and circus cats are out there in the numbers that sightings suggest, it is truly remarkable that so few have been recovered.

Di Francis, author of *Cat Country* (1983), has a different idea about big cats in Britain. He believes that a large, pantherlike cat has survived on the British Isles since Pleistocene times—more than 10,000 years ago! But where is the evidence, asks critic Lena G. Bottriell: "[Not] one skin in a period of some 1,000 years or more? No reports when the population of the same island [England] reduced a smaller cat [*Felis silvestris grampia*] to the point of virtually killing it out?" And like other doubters, she also pointed out the feeding needs of such cats and the serious effect that they would have on the stag and deer population—and on farm animals.

These holes have led others to suggest even stranger theories. One of them is teleportation. Here we are asked to believe that these big cats are real animals that are instantly transported from their native homes to distant places—where they stay for a few days to a few months—before returning in the same way. Other "paranormal," or outside the normal, explanations suggest that the creatures are materialized mental images from the human mind or that they are intruders from a parallel world outside our own (also see entry: **Fourth Dimension**). But what these theories fail to account for is that sightings of big cats almost *never* include supernatural features. The only strange thing about such sightings is that these big cats are in places where they have no right to be.

Big Cats Down Under

Black panthers are still sighted in Australia. Most reports come from the southern coast of New South Wales, but Paul Cropper, who has been investigating the matter for two decades, has collected others from all over the continent.

As with such sightings elsewhere in the world, some have occurred in daylight and at close range. And the animals have been linked to the slaying of livestock—between 1956 and 1957, one Uralla, New South Wales, farmer lost 340 sheep to a large black cat that hunters could not destroy. Prints found at the scenes of sightings usually contained claw marks, but none of the cats have ever been caught or killed. Cropper investigated one report of a killing on the Cambewarra Range in November 1977. When shown the animal's skin, though, he found it to be that of a wild house cat.

While it is tempting to explain all sightings this way—as cases of mistaken identification—the theory does not fit all reports. Consider this sighting, said to have taken place in mid-1975 in the Southern Highlands of New South Wales. Cropper wrote:

> [A farmer] and his son had been out feeding their pigs around 5 o'clock when he … looked up and [saw] a large black animal unhurriedly ambling along a fence, past their sawmill for a couple of hundred yards. They both had watched this creature at a distance of 300 yards for at least 4 or 5 minutes. He estimated the animal weighed between 4 or 5 hundred pounds and stood 2 foot 6 inches high at the shoulder and looked exactly like a black panther. As they watched, the animal sprang 9 ft. to clear a creek, and then disappeared into the bush, leaving a perfect set of pawprints in the soft soil of the creek bank.

The farmer told a neighbor, who had seen the animal earlier, as had his granddaughter in a separate sighting. The two farmers, with several others, went to the creek bank and took a cast of the best preserved of the creature's tracks. They were huge: four inches by five inches.

Far better known is the case of the Queensland tiger. First reported by a number of witnesses around Queensland's Cardwell Bay district, the mysterious animal would be seen in other parts of Australia as well. One early account of the

REEL LIFE

Cat People, 1942.

A young dress designer is the victim of a curse that changes her into a deadly panther who must kill to survive. A classic among the horror genre with unrelenting terror from the beginning to end. Remade in 1982.

creature came from police magistrate Brinsley G. Sheridan, whose 13-year-old son had run across the beast during a walk along the shore of Rockingham Bay. The boy's small terrier had picked up the animal's scent and followed it for roughly half a mile. Then the boy observed: "It was lying camped in the long grass and was as big as a native Dog [dingo; a wild dog found in Australia]; its face was round like that of a cat, it had a long tail, and its body was striped from the ribs under the belly with yellow and black." The animal threw the terrier when it came too close and retreated up a leaning tree. Then it savagely rushed toward the two—and the boy and his dog took off running.

Later witnesses of the Queensland tiger included naturalist George Sharp, who, in the early part of the century, saw a similar creature at twilight along the Tully River. It was, he said, "larger and darker than the Tasmanian Tiger, with the stripes showing very distinctly." Not long afterward a farmer killed such a beast after it had attacked his goats. Sharp followed its tracks through the bush until he came upon its lifeless form. By then wild pigs had eaten most of the head and body, but just enough remained to show that it was about five feet long. Sharp had nothing with which to preserve what was left, though, so it soon rotted away.

Ion L. Idriess, a longtime York Peninsula resident, reported once seeing a "tiger" rip out the insides of a fully grown kangaroo. Another time he found the body of such a creature along the Alice River, where it had died in a fight with his hunting dog, which also lay dead nearby. Idriess described the "tiger-cat" as the size of a "hefty, medium-sized dog. His body is lithe and sleek and beautifully striped in black and grey. His pads are armed with lance-like claws of great tearing strength. His ears are sharp and pricked, and his head is shaped like that of a tiger."

And A. S. Le Souef and H. Burrell wrote of a similar creature in their 1926 book *The Wild Animals of Australasia*. They referred to a "large striped animal which has been aptly described as 'a cat just growing into a tiger.' The animal ... lives in country that man seldom penetrates.... Its stronghold appears to be the rough, rocky country on top of the ranges ... usually covered with heavy forest."

Sightings of this animal, though rare, continue to the present day. If reports of dead specimens are true, then the creature may, in fact, be *real*. While most Australian zoologists find black panther sightings highly questionable, they do not feel the same about reports of the Queensland tiger. As one writer put it, the animal is a "near-candidate for scientific recognition"—not as a big cat, but as the marsupial or pouched "lion" known as *thylacoleo,* which was thought to have died out some 10,000 years ago, having left many fossil remains. It does seem to be the animal that eyewitness accounts describe, even down to the creature's protruding fangs. If this is so, it is perhaps only a matter of time before a living or dead specimen falls into the hands of a zoologist and its existence is proven at last—unless, of course, the species has died out in recent years.

Sources:

Bord, Janet, and Colin Bord, *Alien Animals,* Harrisburg, Pennsylvania: Stackpole Books, 1981.

Clark, Jerome, and Loren Coleman, *Creatures of the Outer Edge,* New York: Warner Books, 1978.

Coleman, Loren, *Mysterious America,* Boston: Faber and Faber, 1983.

Heuvelmans, Bernard, *On the Track of Unknown Animals,* New York: Hill and Wang, 1958.

Shuker, Karl P. N., *Mystery Cats of the World: From Blue Tigers to Exmoor Beasts,* London: Robert Hale, 1989.

BEAST OF EXMOOR

In the spring of 1983 the name "Beast of Exmoor" was given to a mysterious roaming animal that killed a ewe belonging to farmer Eric Ley of South Molton, Devonshire, England. In the next two and a half months, Ley would lose 100 more sheep. The killer beast did not attack its victims at the hindquarters, as a dog or a fox would. Instead, it ripped out their throats.

Most who have seen the Beast of Exmoor describe it as a huge, jet-black cat, eight feet long from nose to tail, though a few witnesses have reported that it is tan-colored. In some cases, two giant cats—one black and one tan—have been seen traveling together! And a small number of witnesses have reported seeing large animals that look like unusual dogs.

First Sightings

Sightings of the "Beast" go back at least to the early 1970s. But it was not until the Ley sheep killings that Britain's Royal Marines were called into the area and London's *Daily Express* offered a one-thousand-pound reward for the beast's death or capture. Marine sharpshooters hid in the hills, and some even said they saw a "black and powerful animal" but were unable to get a clear shot at it. The beast or beasts seemed to quiet their activities while the soldiers were around, but as soon as they left the attacks began again.

One witness, local naturalist Trevor Beer, reported that he saw the beast in the summer of 1984 while watching birds in an area where slaughtered deer had been found. "I saw the head and shoulders of a

The killer beast did not attack its victims at the hindquarters, as a dog or a fox would. Instead, it ripped out their throats.

Black Beast of Exmoor, Somerset, England, photographed by Trevor Beer in 1987.

large animal appear out of the bushes," he wrote. "It looked black and rather otter-like, a first impression I shall always remember for the head was broad and sleek with small ears. The animal's eyes were clear greeny-yellow.... As it stared back at me I could clearly make out the thickish neck, the powerful forelegs and deep chest, and then without a sound it turned and moved swiftly away through the trees. That it was jet black I was sure, and long in the body and tail. I guessed at four and a half feet in body length, and about two feet at the shoulders." Beer chased the animal to the edge of the woods; the way it ran made him think of a "beautiful, very large black panther."

In 1988 an area farmer reported seeing a "fantastic cat going at a hell of a speed. Every time it moved you could see the lights shine back across its ribs." Another time the same man saw a huge cat "jump a hedge, 15 feet from standing, with a fair-sized lamb in its mouth." Late one night in December 1991, a family watched a large pantherlike animal for some minutes as it prowled around their country home. Several weeks earlier the son, 13, had seen it or a similar animal climbing a tree.

An article in London's *Daily Telegraph* reported that by early 1992, a large number of people living in the wild countryside of southwestern England had seen the beast or beasts. There were a variety of theories to explain the sightings: witnesses had mistaken large dogs for the creature, or—more likely—overjudged the size of the animals, which were really house cats gone wild. Other theorists suggested that a small breeding population of pumas, let loose by people who once kept them as pets, roamed England's wild West Country. A more far-fetched idea, held by author Di Francis but rejected by most zoologists, proposed that large cats had secretly lived in Britain since prehistoric times.

Complicating Britain's big cat puzzle is the fact that the creatures have been sighted *all over* the country. Officially, Britain's only native wildcat is *Felis silvestris grampia,* a small feline that lives in the rugged regions of northern England and Scotland (also see entry: **Black Panthers and Other Strange Cats**).

Sources:

Beer, Trevor, *The Beast of Exmoor: Fact or Legend?* Barnstaple, Devonshire, England: Countryside Productions, 1985.
Francis, Di, *Cat Country: The Quest for the British Big Cat,* North Pomfret, Vermont: David and Charles, 1983.
Martin, Andrew, "In the Grip of the Beast," *London Daily Telegraph,* January 4, 1992.

BEAST OF GEVAUDAN

One June day in 1764 in a forest in southeastern France, a young woman tending cows looked up to see a hideous beast bearing down on her. It was the size of a cow or donkey, but it resembled an enormous wolf. The woman's dogs fled, but the cattle drove the animal away with their horns. This cowherd was much more fortunate than most later witnesses of what would come to be known as the "Beast of Gevaudan."

Before long the maimed bodies of shepherd men, women, and especially children became common to the area. The first suspected victim was a little girl who was found in July with her heart ripped from her body. The killings resumed in late August or early September, and soon the creature was fearlessly attacking groups of men. The terrified countryfolk were certain that a *loup-garou* (**werewolf**) was running wild. These rumors gained authority when those who had shot or stabbed the creature reported that it seemed unaffected by their attempts to kill it. On October 8, after two hunters pumped several rifle balls into it from close range, the creature limped off. When word of the incident spread, it was believed, briefly, that the beast had gone off to die. But within a day or two it was killing again!

Terror in the French Countryside

Late in 1764 the Paris Gazette put together a general description of the beast from eyewitness accounts: it was "much higher than a wolf ... and his feet are armed with talons. His hair is reddish, his head large, and the muzzle of it is shaped like that of a greyhound; his ears are small and straight; his breast is wide and gray; his back streaked with black; his large mouth is provided with sharp teeth." On June 6, 1765, the English St. James's Chronicle remarked of the creature, "It appears that he is neither Wolf, Tiger, nor Hyena, but probably a Mongrel, generated between the two last, and forming, as it were, a new Species."

After a frightening public attack on two children, who were bitten and torn even as older youths slashed at the creature with pitchforks and knives, an appeal for help was sent to the Royal Court at Versailles. King Louis XV responded with a troop of horsemen, under the direction of Captain Duhamel. Duhamel ordered several of his men to dress as women on the theory that the creature was especially attracted to females. The soldiers spotted the beast a number of times and shot at it, but it always managed to escape. Finally, after the killings seemed to have stopped, Duhamel assumed that the beast had died of its wounds. After he and his men left, however, the bloodshed resumed.

A large reward for the slaying of the beast brought professional hunters and soldiers to the area. More than 100 wolves were killed, but the creature's rampage continued. Some hunters, including a professional wolf-tracker who had been sent personally by the king, reported that they had badly wounded the beast. But nothing seemed to stop it. During the summer of 1765 the massacre of children was especially fierce.

As the months dragged on, whole villages were abandoned after residents claimed they had seen the beast staring through their windows. Those who ventured out into the streets were attacked. Many peasants were too petrified to fire on the creature even when they had an open shot.

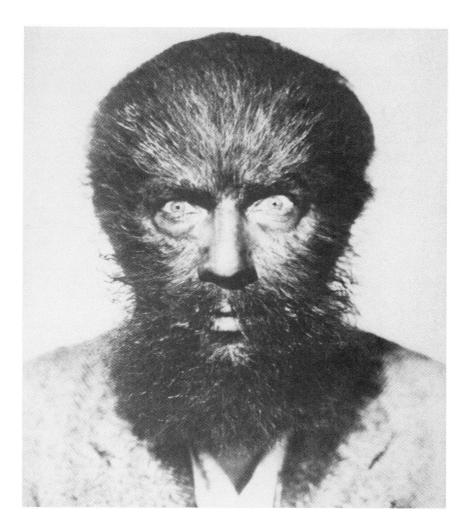

Close-up of wolf-man from 1932 film *Island of Lost Souls.* Many of the country-folk terrorized by the Beast of Gevaudan believed it to be a form of the legendary werewolf.

Death by a Silver Bullet

The crisis was finally resolved in June 1767. The Marquis d'Apcher, who lived in the western part of Gevaudan, marshaled several hundred hunters and trackers who fanned out in smaller bands over the countryside. On the evening of the 19th, the beast charged members of one group. Jean Chastel, who had loaded his weapon with *silver* bullets because the creature was so widely rumored to be a werewolf, fired on it twice. The second shot hit it squarely in the heart and killed it. The collarbone of a young girl was found in the animal's stomach when it was cut open. By the time of its death, the beast had killed some 60 people.

By the time of its death, the beast had killed some 60 people.

After the monstrous carcass was paraded throughout the region for two weeks, it was packed up to be sent to Versailles. By that time, though, it had begun to rot. Before it reached the king's court, it had to be buried somewhere in the French countryside.

Modern wildlife experts doubt reports of wolves attacking people, maintaining that the animals try to stay clear of humans. Yet there are widespread and seemingly believable reports of man-eating wolves, especially in the days before guns. Folklorists W. M. S. and Claire Russell point out that "modern wolves have had many generations' experience of fire-arms, and are likely to be more cautious than their ancestors." The story of the Beast of Gevaudan demonstrates extraordinary animal behavior. And the sheer size of the creature makes one wonder if it really was a wolf. Perhaps it was a fierce and unknown wolflike creature.

Sources:

Caras, Roger A., *Dangerous to Man: The Definitive Story of Wildlife's Reputed Dangers,* New York: Holt, Rinehart and Winston, 1975.

Russell, W. M. S., and Claire Russell, "The Social Biology of Werewolves," *Animals in Folklore,* edited by J. R. Porter and W. M. S. Russell, Totowa, New Jersey: Rowman and Littlefield, 1978.

ALLIGATORS IN SEWERS

There is a persistent American legend that alligators live under New York City, in its sewer system. Supposedly, tiny pet alligators were flushed down toilets when their owners grew tired of them—and they survived, growing so large that they became a threat to sewer workers! New York City officials today deny that any such creatures exist.

Though the rumor was most widespread in the 1960s, it was actually based on real and puzzling events that took place in the 1930s. The first of these occurred on June 28, 1932, when "swarms" of alligators were seen in the Bronx River. A three-foot-long specimen was found dead along its banks. In March 1935 and June 1937 both live and dead alligators were discovered.

Perhaps the most remarkable event was the one reported in the *New York Times* of February 10, 1935. Several teenage boys were shov-

Alligators and crocodiles have a history of showing up in the most unexpected places!

eling snow into an open manhole near the Harlem River when they spotted something moving in the icy water ten feet below. It turned out to be an alligator trying to free itself. The boys got a rope, twisted it into a lasso, and pulled the animal to the surface. When one of them tried to take the rope off the alligator's neck, however, it snapped at him. Feeling threatened, the young men beat it to death with their shovels.

The boys dragged the dead creature to a nearby auto repair shop, where the animal was weighed and measured. It was 125 pounds and seven and a half feet long! The police were informed, and a city sanitation worker took the body of the alligator off to be burned.

Around that time, Teddy May, New York City's superintendent of sewers, was receiving complaints from his workers about alligators. At first he thought the reports were just the imaginings of those who were secretly drinking alcohol on the job, and he even hired investigators to observe the habits of his employees. When these investigators came

It was 125 pounds and seven and a half feet long!

back with nothing to report, however, May himself went into the sewers with a flashlight—which soon enough revealed the presence of alligators! Shaken, May had the animals destroyed by poison or gunshot.

It was not clear how the animals got there, though it was generally thought that they were unwanted or escaped pets.

According to animal expert Loren Coleman, who specializes in these strange appearances, between 1843 and 1983 some 84 such animals were either sighted or recovered dead or alive across the United States and Canada. Coleman believed that the "pet escapee explanations cannot deal adequately with these accounts of alligators in northern waters—when it is caimans [animals similar to alligators from Central or South America] that are sold as pets." For caimans live in tropical waters and could not survive long in harsh northern climates. But alligators and crocodiles do have a history of showing up in the most unexpected places, and the mystery of these displaced animals has never been satisfactorily resolved.

Sources:

"Alligator Found in Uptown Sewer," *New York Times,* February 10, 1935.
Coleman, Loren, "Alligators-in-the-Sewers: A Journalistic Origin," *Journal of American Folklore,* July/September 1979, pp. 335-338.
Coleman, Loren, *Mysterious America,* Boston: Faber and Faber, 1983.
Michell, John, and Robert J. M. Rickard, *Living Wonders: Mysteries and Curiosities of the Animal World,* London: Thames and Hudson, 1982.

ENTOMBED ANIMALS

In 1890 a writer for *Scientific American* noted that "many well authenticated stories of the finding of live toads and frogs in solid rock are on record." Twenty years later a *Nature* editor forcefully disagreed, stating, "The thing is absolutely impossible, and ... our believing it would involve the conclusion that the whole science of geology (not [to] speak of biology also) is a mass of nonsense." Because of their outrageousness, accounts of entombed toads, frogs, and other animals are seldom discussed in the scientific literature of our time. But they were reported frequently in respected journals of the nineteenth and earlier centuries.

Loads of Toads, Frogs, and Lizards

One such account was related by Ambroise Pare, chief surgeon to Henry III of France, during the sixteenth century. It was reprinted in the 1761 edition of the *Annual Register*. While watching a quarryman "break some very large and hard stones, in the middle of one we found a huge toad, full of life and without any visible aperture [opening] by which it could get there," Pare stated. "The laborer told me it was not the first time he had met with a toad and the like creatures within huge blocks of stone."

On April 7, 1865, laborers at Hartlepool, England, made a discovery that caused quite a stir. While breaking up a block of magnesium limestone 25 feet underground, they split open a cavity in which, to their astonishment, was a living toad. "The cavity was no larger than its body, and presented the appearance of being a cast of it," the *Hartlepool Free Press* reported on April 15. "The toad's eyes shone with unusual brilliancy, and it was full of vivacity on its liberation."

At first the toad seemed to have some difficulty breathing, and a "barking" sound came out of its nostrils. This soon stopped, however, though the creature still gave a startled bark whenever it was touched. When discovered, the toad was of a pale color that matched its stony prison. But in a short time it grew darker until it became olive-brown. "The claws of its fore feet are turned inwards," the newspaper noted, "and its hind ones are of extraordinary length and unlike the present English toad."

Lumbermen also reported finding living toads embedded in or tumbling out of the solid trunks of trees that they were sawing. In 1719 the *Memoires* of the French Academy of Sciences related that a medium-sized live toad was found "in the foot of an elm, ... three or four feet above the root and exactly in the center, ... filling up the whole vacant space." In the fall of 1876, South African timbermen cutting a 16-foot trunk into lumber had just removed the bark and the first plank when a hole the size of a wine glass was uncovered. Inside this space were 68 small toads, each about the size of the tip of a human finger! According to the *Uitenhage Times* (December 10), "They were ... of a light brown, almost yellow color, and perfectly healthy, hopping about and away as if nothing had happened. All about them was solid yellow wood, with nothing to indicate how they could have got there, how long they had been there, or how they could have lived without food, drink, or air."

Next to toads, frogs are the most frequently entombed animals. Celebrated twentieth-century biologist-philosopher Sir Julian Huxley

> While breaking up a block of magnesium limestone 25 feet underground, they split open a cavity in which, to their astonishment, was a living toad.

"The toad's eyes shone with unusual brilliancy, and it was full of vivacity on its liberation."

related this story, told to him in a letter from Eric G. Mackley, a gas fitter from Barnstaple, Devonshire, England. Mackley explained that in the midst of a road expansion, he and a fellow worker had to move gas meters located inside the front gates of a row of bungalows. The meter-houses had concrete floors, which had to be broken up for pipe extensions. "My mate was at work with a sledge hammer when he dropped it suddenly and said, 'That looks like a frog's leg,'" Mackley recalled. "We both bent down and there was the frog.... [The] sledge was set aside and I cut the rest of the block carefully. We released 23 perfectly formed but minute frogs which all hopped away to the flower garden."

Tilloch's Philosophical Magazine offered a lizard-in-stone tale in 1821. A Scottish mason, David Virtue, was fashioning "a barley mill-stone from a large block" when "he found a lizard imbedded in the stone. It was about an inch and a quarter long, of a brownish yellow color, and had a round head, with bright sparkling projecting eyes. It was apparently dead, but after being about five minutes exposed to the air it showed signs of life.... It soon after ran about with much celerity." Virtue noted that when the lizard was first discovered, "it was coiled up in a round cavity of its own form, being an exact impression of the animal," and that the stone and creature were of the same color. The block from which the animal sprang had come from 21 or 22 feet below the earth's surface! The mason remarked that "the stone had no fissure [and] was quite hard."

Explanations

Like all mysteries, entombed animal stories have stirred both excitement and disbelief. The *Nature* writer quoted earlier offered this explanation for the phenomenon: "The true interpretation of these alleged occurrences appears to be simply this—a frog or toad is hopping about while a stone is being broken, and the non-scientific observer immediately rushes to the conclusion that he has seen the creature dropping out of the stone itself."

But this explanation is not consistent with such reports; it does not take into account the presence of a "smooth" or "polished" cavity—only slightly larger than the creature's body—inside the rock, concrete, or tree. And often the animal is spotted within that cavity before freeing itself or being freed from it. Also, the beast in question frequently boasts an unusual appearance that suggests it has been confined for some length of time.

In a modern case reported in August 1975, construction workers in Forth Worth, Texas, were startled to find a living green turtle in concrete they were breaking up; the surface had been laid more than a year before. The smooth, body-shaped cavity in which the turtle rested during its imprisonment was clearly visible. The animal's good fortune, however, was short-lived; it died within 96 hours of its rescue.

It has been suggested that animals trapped inside rocks have been able to survive by drinking water seeping through cracks. But many witnesses have claimed that no such openings could be observed. What is more, this does not explain how the animals got there in the first place.

Sources:

Michell, John, and Robert J. M. Rickard, *Mysteries and Curiosities of the Animal World,* New York: Thames and Hudson, 1982.
Michell, John, and Robert J. M. Rickard, *Phenomena: A Book of Wonders,* New York: Pantheon Books, 1972.
"Toads in Rocks," *Scientific American* 63, 1890, p. 180.

NORTH AMERICAN APES

Loren Coleman, an expert in strange animal sightings, believes that a population of apes has existed for some time in the dense forests of the Mississippi Valley and surrounding areas. He calls these animals North American apes, or NAPEs.

He bases his idea on a number of reports from mostly southeastern states of "ape-like, hairy, and tailless" creatures. Some are referred to in folklore: for example, early settlers spoke of a tribe of "monkeys" that lived in the surrounding woods of Monkey Cave Hollow near Scottsville, Kentucky. Other evidence includes twentieth-century reports of contacts with "gorillas" and "chimpanzees" in the North American wild.

Though these animals are often viewed as escapees from circuses or zoos, in fact such escapes are rare. In the 1970s officials admitted that small populations of wild primates did exist in Florida and Texas. But Coleman believes that NAPEs are something else; he thinks that they are descendants of *Dryopithecus,* a widespread great ape about the size of a chimpanzee. The dryopithecines were supposed to have died out during Pleistocene times, more than 10,000 years ago. According to Coleman, the animals remain in North America and, moreover, have learned how to swim. He feels that "their range up and down the Mississippi" and its branches demonstrates that they travel by water as well as through the "forests bordering the river systems."

Though sightings of NAPEs have brought doubts and varying explanations, Coleman's best evidence consists of footprints that seem

Coleman's best evidence consists of footprints that seem to show the opposed left toe of a chimpanzee or, possibly, a lowland gorilla.

to show the opposed left toe of a chimpanzee or, possibly, a lowland gorilla. Coleman's interest in the subject began, in fact, with his own discovery of such a print along a dry creek bed near Decatur, Illinois, in the spring of 1962. Similar prints have been found in Florida, Alabama, and Oklahoma.

Most anthropologists and primatologists have paid no attention to Coleman's rather wild theory. Roderick Sprague of the University of Idaho, however, has praised Coleman's efforts, although he admitted that Coleman's idea of a modern *Dryopithecus*—without any fossil evidence—"is a weak point in an otherwise well developed argument" for the existence of apes in North America.

Sources:

Coleman, Loren, *Mysterious America,* Boston: Faber and Faber, 1983.
Sanderson, Ivan T., *Abominable Snowmen: Legend Come to Life,* Philadelphia: Chilton Books, 1961.

Loren Coleman found this apelike print in a creek bed in Decatur, Illinois, in 1962.

Shaggy, Two-footed Creatures in North America

- HAIRY BIPEDS

- BIGFOOT

- MOMO

- LAKE WORTH MONSTER

- JACKO

- MINNESOTA ICEMAN

Shaggy, Two-footed Creatures in North America

HAIRY BIPEDS

One night in January 1992, two men driving on a dark country road were startled when their headlights made out two figures. The larger figure stood seven to eight feet tall and appeared to weigh over 500 pounds; the shorter one looked roughly five feet and 300 pounds. The creatures were moving toward the car. Frightened, the driver backed up until he found a place to turn around. Looking over their shoulders as they sped away, the witnesses saw that the bigger creature was still heading in their direction.

This story sounds like it should originate in the Pacific Northwest, the reported home of **Bigfoot** (also known as Sasquatch), the giant apelike human or humanlike ape that many think exists but has managed to escape scientific detection. But the above report comes from Tuscola County in eastern Michigan. According to local monster expert Wayne King, it is the county's 38th hairy biped (an animal that walks on two feet) report since 1977. In fact, similar sightings have been recorded in nearly every state and province in the United States and Canada.

While they seem similar, there is a difference between Bigfoot reports and accounts involving hairy bipeds (HBs). Bigfoot sightings and the physical evidence that goes with them (mostly footprints) do not shake our ideas about reality; they actually adhere to the laws of zoology. In

> ### PRIMATOLOGY AND ANTHROPOLOGY
>
> The order of Primates includes man, apes, monkeys, and related forms. Scientists who study primates are called *primatologists*. *Anthropologists* study human beings. They focus on the physical, social, and cultural development and behavior of men and women.

In the 1945 film *White Pongo,* this mythic white gorilla is believed to be the "missing link."

fact, several well-respected anthropologists and primatologists, especially Grover Krantz and John Napier, have written about the creature's possible place in the families of ape and man (also see entry: **Bigfoot**). But sightings of HBs can be out of this world, as we shall see.

Nineteenth- and Twentieth-Century Hairy Biped Reports

The mystery of Bigfoot grabbed worldwide attention in the late 1950s. Therefore, it might seem that Bigfoot's popularity inspired those with active imaginations to encounter their own HBs. But Americans were reporting HBs long before they had ever heard of Bigfoot. For example, in the early 1970s, while teaching at Newfoundland's Memorial University, American folklore expert David J. Hufford found a late-nineteenth-century book that recorded sightings of the "Traverspine gorilla," a beast that derived its name from the settlement around which it was often seen.

Ivan T. Sanderson

(1911-1973)

Ivan Terence Sanderson was a natural scientist known for several popular books—*Animal Treasures, Animals Nobody Knows,* and *Living Mammals of the World.* During his career he appeared often on television, presenting exotic animals to audiences.

Born in Edinburgh, Scotland, he spent part of his childhood on his father's game preserve in Kenya. Sanderson later earned master's degrees in zoology, botany, and geology. After launching a number of scientific expeditions in remote locales around the world in the 1930s, he began a full-time career as a writer and lecturer on the natural sciences.

Sanderson's focus on unexplained natural events brought cryptozoology to the attention of the public. In 1948 he wrote a popular article about still-living dinosaurs in Africa. In 1961 he published *Abominable Snowmen: Legend Come to Life,* the first book of its kind to explore worldwide reports of apeman-like creatures. Sanderson also wrote about everything from UFOs to sky-born falling rocks to the Bermuda Triangle. In 1965 he formed the Society for the Investigation of the Unexplained (SITU).

In the late 1960s Sanderson wrote two books with unusual theories about UFOs. *Uninvited Visitors* (1967) argued that UFOs were atmospheric life forms (also see entry: Star Jelly). *Invisible Residents* argued that some UFOs were piloted by an advanced underwater civilization that has always lived side by side with us (also see entry: Vile Vortices). Sanderson died of cancer in 1973.

Sightings in the early twentieth century include a *Washington Post* report of a "huge gorilla" wandering in the woods near Elizabeth, Illinois, on July 25, 1929. In *Wild Talents* (1932) strange-animal expert Charles Fort wrote about a hunt for an "apelike animal—hairy creature, about four feet tall" that went on for three weeks in June and July 1931. Local circus and zoo spokesmen insisted that none of their apes were missing. The police were called in, a "gorilla" was spotted several times, and tracks were found that "seemed to be solely of those of the hind feet." And the following January, in a country area north of Downington, Pennsylvania, John McCandless heard a moaning sound in a bush. Its source, he told a reporter, was a "hideous form, half-man, half-beast, on all fours and covered with dirt or hair." Soon afterward other people told of coming upon the creature, which managed to dodge search parties.

Between the 1920s and the 1950s, other reports of HBs were recorded in New Jersey, Maryland, Missouri, Indiana, Michigan, Alabama, and other states. Later, during the Bigfoot era (from the late 1950s to the present), a number of people came forward to describe sightings from earlier in the century. A woman told biologist Ivan T. Sanderson—the first writer to bring Bigfoot to wide public attention—about a 1911 event that took place when she was living in far northern Minnesota; there, she said, two hunters saw a "human giant which had long arms and short, light hair" and left strange prints. A man remembered that in 1942, while he was cutting trees in a New Hampshire forest, a "gorilla-looking" creature followed him for some 20 minutes. And in 1914, according to an account given in 1975, a boy saw a gorilla-like creature sitting on a log in his backyard in Churchville, Maryland.

Later HB Sightings

From the 1950s to the present, HB reports have been recorded in startling numbers. Here are a few examples:

Monroe, Michigan, August 11, 1965

As they rounded a curve in a wooded area, Christine Van Acker, 17, and her mother were astonished as a hairy giant stepped out onto the road. In her panic, Christine hit the brakes instead of the gas. As she frantically tried to restart the car, the creature—seven feet tall and emitting a rank odor—reached through the open window and grabbed the top of her head.

The screams of Christine and her mother and the honking car horn caused the HB to retreat to the woods. Nearby workmen arrived at the scene moments later, finding the two women dazed with fear. Somehow in the course of the attack, Christine had received a black eye. The story received national attention, with a photograph of Christine's bruised face appearing in hundreds of newspapers around the country.

Rising Sun, Indiana, May 19, 1969

At 7:30 P.M., as George Kaiser was crossing the farmyard on the way to his tractor, he spotted a strange figure standing 25 feet away. He was able to watch it for about two minutes before it spotted him. Kaiser described the creature as "upright," "about five-eight or so," and "very muscular." He added, "The head sat directly on the shoulders, and the face was black.... It had eyes set close together, and with a very short forehead. It was covered with hair except for the back of the hands and the face. The hands looked like normal hands, not claws." Seeing Kaiser, the creature made a grunting sound, turned around, leaped over a ditch, and dashed off at great speed down the road. Plaster casts of the tracks it left showed three small toes plus a big toe.

Putnam County, Indiana, August 1972

Randy and Lou Rogers, a young couple living outside tiny Roachdale (population 950), 40 miles west of Indianapolis, experienced regular late-night visits from a shadowy creature. Brief glimpses revealed a large, hairy "gorilla." It traveled on two feet most of the time, but when it ran, it did so on all fours. Mrs. Rogers reported the odd observation that "we could never find tracks, even when it ran over mud. It would run and jump, but it was like somehow it wasn't touching anything. When it ran through weeds, you couldn't hear anything. And sometimes when you looked at it, it seemed you could see through it."

Regardless, local farmer Carter Burdine claimed to have lost 30 of his 200 chickens to the creature, which tore them apart. Burdine, his father, and his uncle saw the HB in the chicken house and chased it into the barn. The uncle opened fire on it as it fled to a nearby field. "I shot four times with a pump shotgun," Bill Burdine said. "The thing was only about 100 feet away when I started shooting. I must have hit it. I've killed a lot of rabbits at that distance." Still, the HB appeared unharmed! At least 40 other persons claimed to have seen the creature before sightings suddenly stopped later in the month.

Vaughn, Montana, December 26, 1975

In the late afternoon, two teenaged girls went to check on their horses, which seemed upset. Two hundred yards from them and 25 yards from a thicket, they observed a huge figure, seven and a half feet tall and twice as wide as a man. Intending to frighten it off, one of the girls fired a .22 rifle into the air. When nothing happened, she fired again, and this time the creature dropped to all fours, walked a short distance, then stood up again. The girls took off running. As one looked over her shoulder, she saw three or four similar creatures with the first beast, all heading toward the thicket. Law officers asked the girls to take a lie-detector test, which they passed. Other sightings, hearings, and tracks of HBs were recorded in Montana in the mid-1970s.

Salisbury, New Hampshire, October 1987

Two or three days after a hunter had told him of seeing two strange beasts walking across a field next to Mill Brook, Walter Bowers, Sr., sensed that he was being watched while hunting in the area. Then, standing between two groups of trees, he saw a "thing ... at least nine feet" tall, "maybe less, maybe more"—a creature covered with grayish-colored hair. Because the sun was in his eyes, Bowers could not make out the beast's face, but he noted that the "hands were like yours or mine, only three times bigger.... It was just like ... a gorilla, but this here wasn't a gorilla." The HB ran into a swamp, and Bowers ran to his car and sped away. A reporter described the witness as a "man of sound mind and sober spirit." Nonetheless, a game warden suggested that the man had merely spotted a moose.

Furry Objects and Flying Objects

The HB sightings in Roachdale (Putnam County), Indiana, were linked to another strange occurrence. Late one August evening in 1972, a witness in Roachdale saw a glowing object hover briefly over a cornfield before it seemed to "blow up." Lou Rogers, who lived on the other side of the field, observed an HB soon afterward.

While the witness may have just seen an exploding meteor, unconnected to the sighting, a handful of cases do link HBs and **UFOs** (unidentified flying objects). Some investigators, especially Stan Gordon and Don Worley, have suggested that HBs are a kind of UFO visitor. Though most of their cases are poorly documented—usually no more than a UFO sighting in the same general area as an HB report—there have been a few incidents that have proven startling. For example:

Uniontown, Pennsylvania, October 25, 1973

Having observed a red light hovering above a field just outside town, a 22-year-old man and two 10-year-old boys rushed to the site in a pickup truck. There they saw a white dome-shaped UFO resting on the ground, "making a sound like a lawn mower." "Screaming sounds" could be heard nearby. Two large apelike creatures with glowing green eyes were walking along a fence. The taller, eight-foot HB was running its left hand along the fence, while the other hand nearly touched the ground; behind it, a shorter, seven-foot creature was trying to keep up. The whining sound they both made seemed to be their way of communicating. The eldest witness, who had a rifle, fired three times

directly into the larger HB, which whined and reached out for its companion. At that moment the UFO vanished and the two creatures disappeared into the trees. A state police officer called to the scene soon afterward noticed a 150-foot glowing area where the UFO had sat. He also heard loud crashing sounds in the woods, made by someone or something big and heavy. At this point, the 22-year-old witness experienced a nervous collapse.

More Strangeness

On March 28, 1987, at 11:45 P.M., Dan Masias of Green Mountain Falls, Colorado, happened to look out his window to see "these creatures ... running down the road right in front of my house, which at one point is 30 feet from my front window. The whole road there was covered with about a quarter of an inch of fresh, cold snow that had fallen. They ran down the road in a manner with their arms hanging down, swinging in a pendulum motion. The first impression I got was that they were covered with hair. It was the most incredible thing I've ever seen."

After Masias's sighting was reported in the newspapers, other residents of the area, near Pike National Forest, came forward with their own accounts (which they had kept quiet for fear they'd be mocked). Sightings and hearings—of unearthly howls and growls—continued, and some people who followed HB tracks in snow swore that they vanished in midstep.

Just as the idea that apelike creatures could secretly live in states like Indiana and New Hampshire seems biologically impossible, the very nature of HBs seem biologically impossible as well. Sometimes they don't leave tracks when it seems that they should; and when they do leave tracks, they may be two-, three-, four-, five-, or even six-toed! In a handful of accounts, we are told that HBs have been shot and killed, but more often witnesses insist that the bullets barely bother the creatures. Sometimes HBs seems to disappear instantly like ghosts, but they leave things—like strands of hair caught in fences—behind.

Even more incredible, some witnesses, in locations as far apart as southern California and South Dakota, have reported *invisible* HBs! During a flurry of sightings at a Native American reservation in South Dakota in 1977, a creature was seen throughout the afternoon and evening hours on November 3. By then local residents and law officers had the area staked out. One of them, rancher Lyle Maxon, reported this weird event:

> We were out there walking in the dark, and I could hear very plainly something out of breath from running.... I put my flashlight right where I could plainly hear it, only where it

REEL LIFE

King Kong, 1933.

The original beauty and the beast film classic tells the story of Kong, a giant ape captured in Africa and brought to New York as a sideshow attraction. Kong falls for Fay Wray, escapes from his captors, and rampages through the city. Remade numerous times.

The Planet of the Apes, 1968.

Astronauts crash land on a planet where apes are masters and humans are merely brute animals. Charlton Heston delivers a believable performance and superb ape makeup creates realistic pseudo-simians of Roddy MacDowall and other stars. Followed by four sequels and two television series.

Return of the Ape Man 1944.

Campy fun with John Carradine and Bela Lugosi. A mad scientist transplants Carradine's brain into the body of the "missing link."

Scene from 1933 horror classic *King Kong.*

should have been, there was nothing in sight. Now what I'm wondering is, can this thing make itself invisible when things get too close for comfort?

In *Bigfoot,* a 1976 book on HB sightings in southern California, B. Ann Slate and Alan Berry tell of similar events.

It is certain that at least some HB reports are hoaxes—stories told by liars or by people who were truly fooled by pranksters wearing costumes. Other HBs (for example, a figure observed by a number of Lawton, Oklahoma, residents in February 1971) turn out to be dirty, bearded, mentally ill humans like the "wild men" of so many nineteenth-century accounts. And in some cases HBs are probably bears. Still, these explanations can't possibly account for all the HB sightings—and there are now many—that have been reported. So far no general theory has been suggested to explain these puzzling reports.

Sources:

Bord, Janet, and Colin Bord, *Alien Animals,* Harrisburg, Pennsylvania: Stackpole Books, 1981.

Brandon, Craig, "Bigfoot!!! Apeman Stalks Woods Near Us, Book Claims,"*Albany* [New York] *Times Union,* February 16, 1992.

Drier, Mary, "Bigfoot Returns, Area Men Maintain, *Bay City* [Michigan] *Times,* March 24, 1992.

Sanderson, Ivan T., *Things,* New York: Pyramid Books, 1967.

Slate, B. Ann, and Alan Berry, *Bigfoot,* New York: Bantam Books, 1976.

BIGFOOT

Bigfoot is, without a doubt, North America's most prominent cryptozoological (the science of hidden animals) mystery. If its existence is ever proven (and nothing short of an actual specimen would satisfy most scientists) it would provide an amazing new look into human evolution. For if Bigfoot is out there, it may be a relative of ours. In fact, Bigfoot believers think that it is either an ape or a kind of early human being.

Bigfoot, or Sasquatch, as it is known in Canada, is the giant manlike ape or apelike man reported in the northwestern United States (northern California, Oregon, Washington, and Idaho) and far western Canada (British Columbia and Alberta). This region of mountains and forests is so vast that the idea that such a beast could survive—undetected by all but a few startled eyewitnesses—is at least possible, if still incredible.

According to John Green, a leading Canadian Bigfoot investigator and writer, the creatures average seven and a half feet in height. (Bigfoot eyewitness accounts number in the many hundreds.) They seem to prefer being alone and are seldom seen in the company of others. Hair covers almost all of their bodies, and the length of their limbs is more human than apelike. Nonetheless, their broad shoulders, lack of a neck, flat faces and noses, sloped foreheads, pronounced eyebrow ridges, and cone-shaped heads are more animal-like than human. They eat both animals and plants, are largely nocturnal (active at night), and are mostly inactive during cold weather.

John Napier, a primatologist (one who studies primates, an order of mammals made up of humans and apes), notes that in a number of the most reliable reports, the Sasquatch is "covered in reddish-brown or auburn hair." Still, black, beige, white, and silvery-white hair have also been noted. Napier added that footprints range in size from 12 to 22

The word "Sasquatch" comes from a term used by the Coast Salish Indians.

Noted Sasquatch investigator John Green (left) with two companions and Bigfoot casts.

inches, with the average sample measuring 16 inches in length and 7 inches across.

Background

Reports of a creature resembling Bigfoot began with early American Indian legends of giant woodland bipeds (animals that walk on two feet). The closest creature to Bigfoot in this folklore is the Witiko (or Wendigo), known to the Algonkian Indians of the northern forests. But Witikos were cannibalistic giants with supernatural powers; they could possess people and turn them into other Witikos, for instance.

Published reports of Bigfoot references started to appear in the early twentieth century. The Victoria, British Columbia, newspaper the *Colonist,* in 1901 related the experience of Mike King, a lumberman working on Vancouver Island near Campbell River. King was alone because his packers refused to go with him, fearing the "monkey men" they said lived in the forest. Late in the afternoon the lumberman saw a "man beast" washing roots in the water. When the creature became aware of him, it cried out and scooted up a hill, stopping at one point to look over its shoulder at him. King described the beast as "covered with reddish brown hair, and his arms were peculiarly long and were used freely in climbing and in brush running; the trail showed a distinct human foot, but with phenomenally long and spreading toes."

Three years later, on December 14, 1904, the *Colonist* reported that four reliable witnesses had also seen a Bigfoot-like creature on Vancouver Island. And three years after that, it told of an American Indian village that was abandoned, its inhabitants frightened into moving by a "monkey-like wild man who appears on the beach at night [and] ... howls in an unearthly fashion."

Residents of western Canada were well aware of the hairy giants called Sasquatches. This was because of the writings of J. W. Burns, a schoolteacher at the Chehalis Indian Reserve near Harrison Hot Springs, British Columbia. From the Chehalis Indians, Burns learned that Sasquatch was not so much an apeman as a fabulous superman, an intelligent "giant Indian" with supernatural powers.

Other people who claimed that they had really seen the hairy giants included a British Columbia man named Albert Ostman, who came forward in 1957 to report an incident he claimed took place in 1924. While on a prospecting trip at the head of Toba Inlet, opposite Vancouver Island, he was scooped up one night inside his sleeping bag and carried many miles. When he was finally dumped out he discovered that he was the prisoner of a family—adult male and female, young male and female—of giant apelike creatures! Though the beasts were friendly, they clearly did not want him to escape, and Ostman managed to do so only after the larger male choked on his chewing tobacco. He was gone for six days. Those who interviewed the man, including strange-animal experts John Green and Ivan T. Sanderson, did not doubt his honesty or his mental health. Even primatologist Napier found the account "convincing."

Another fascinating story tells of an attack by Bigfoot creatures on a group of miners in the Mount St. Helens/Lewis River area of southwestern Washington. The event began one evening in July 1924, when two of the miners—already jumpy from listening to a week's worth of strange whistling and thumping sounds coming from nearby ridges—spotted a seven-foot-tall apelike creature and fired on it. They fled to their cabin and, along with two other men, endured a night-long attack by a number of the creatures, who threw rocks and repeatedly tried to smash the door

While on a prospecting trip at the head of Toba Inlet, opposite Vancouver Island, a man was scooped up one night inside his sleeping bag and carried many miles. When he was finally dumped out he discovered that he was the prisoner of a family of giant apelike creatures!

in. *Portland Oregonian* reporters who came to the scene later found giant footprints. The spot where the episode occurred was thenceforth called Ape Canyon.

In 1967 one of the men involved in the incident, Fred Beck, published a booklet with his son Ronald recalling the event; it was titled *I Fought the Apemen of Mt. St. Helens.* Later, however, in a 1982 interview with a Vancouver newspaper, Rant Mullens, 86, confessed that he and his uncle were responsible for the long-ago episode. On their way home from a fishing trip, they "rolled some rocks down" near the cabin as a joke. Ronald Beck rejected the idea that the complicated story of the attack was "a common hoax."

Sasquatch sightings continued and were reported from time to time, mostly in Canadian newspapers. At some point in the 1920s the name "Bigfoot" came into use, as northwestern residents marveled at the size of the tracks they were encountering in remote areas. In 1958 the mystery of Bigfoot grabbed American attention when heavy-equipment operators near Willow Creek in northwestern California discovered a large number of tracks left by a huge biped, which had seemingly examined a land-clearing bulldozer left at the site overnight. After the tracks appeared several more times, casts were made, bringing wide press coverage. A few weeks later, in late October, two men driving down a wilderness road saw a huge hairy biped cross in front of them and disappear into the trees, leaving prints behind.

The Patterson Film

By the 1960s Bigfoot, sometimes called "America's abominable snowman," held a firm place in the public's imagination. Though scientists refused to believe that witnesses were actually seeing what they claimed (reports were blamed either on hoaxes or bear sightings), several investigators, such as John Green, Rene Dahinden, and Jim McClarin, looked for witnesses, went into the bush in search of sightings themselves, studied the data, and wrote articles and books about their findings. Ivan T. Sanderson's 1961 *Abominable Snowmen: Legend Come to Life*—the first book to fully discuss the Bigfoot/Sasquatch phenomenon—linked the North American reports with worldwide accounts of "wild men," including the **Almas** of Mongolia and the **yeti,** or "abominable snowman," of the Himalayas.

Among those who went looking for Bigfoot in the wild was Roger Patterson, a onetime rodeo rider and sometime inventor and promoter. After a 1959 *True* magazine article about Bigfoot sparked his interest, he roamed the Pacific Northwest woods when time permitted, hoping to catch a glimpse of the creature. Eventually he decided to make a documentary film about the mystery and began taking a motion-picture camera with him on his expeditions. He shot footage that he thought might prove useful in his future movie.

At a little after 1:15 on the afternoon of October 20, 1967, Patterson and a companion, Bob Gimlin, were riding north up the partly dry, 100-yard-wide bed of Bluff Creek in the Six Rivers National Forest of northern California. (This area had seen so much Bigfoot activity—both sightings and tracks—that it had become something of a tourist attraction.) At one point a high pile of logs in the center of the stream blocked their way, and they had to direct their horses around it. As they passed the logs and took up their original course, they saw—or would claim to have seen—something that has become the source of heated argument for decades.

Bob Gimlin with prints from the Patterson film site.

They saw a female Bigfoot stand up from the creek water in which she was squatting and walk briskly away into the surrounding trees, swinging her arms all the while. This caused the horses that Patterson and Gimlin were riding (as well as their packhorse) to panic. Patterson's animal reared up and fell over sideways on its rider's right leg. As his horse staggered to its feet, Patterson felt around for the 16-mm camera in the saddlebag, then jumped off to follow the retreating creature on foot. Only 28 feet of film remained in the camera, and Patterson used it to record the Bigfoot from three different positions.

Patterson died in 1972, swearing to the end that both the sighting and the film were real. Gimlin, still alive, also sticks by the story. The first investigator on the site, Bob Titmus, found tracks in locations that exactly matched the Bigfoot's route in the film; he made casts of ten of them. The prints also indicated that the creature had gone up a hillside and sat down for a while, apparently to watch Patterson and Gimlin, who had stopped their tracking efforts in order to recover two of their horses.

Of course Patterson's film did not solve the mystery of whether an "abominable snowman" of America existed or not. As a matter of fact, debate over the Patterson film goes on and will probably continue to

Photograph by Roger Patterson taken northeast of Eureka, California: evidence of Bigfoot or a "man in a suit"?

until the "man in the suit" confesses or someone produces a physical specimen of Bigfoot to compare with the figure in the film. Everyone agrees, at least, that the Patterson footage is worth discussing, unlike many other Bigfoot films that are clearly fake.

More Recent Sightings

The next big stir in the Bigfoot mystery took place in 1982. Though it first looked like a promising development, it would ultimately prove a disappointment.

The event began with a story told by Paul Freeman, a sometimes employee of the U.S. Forest Service. On the morning of June 10, he was driving through the Blue Mountains in the Umatilla National Forest, which stretches across southeastern Washington and northeastern Oregon. Spotting some elk, he stopped his truck, jumped out, and followed the animals on foot. He wanted to find out if there were any calves among them.

As he rounded a bend, he noticed a "stench" and at the other side of the turn saw something coming down a bank though some thick plant growth. When the figure stepped into the clearing, Freeman froze and stared in disbelief at an "enormous creature"—an eight-and-a-half-foot Bigfoot—which stared back at him. For a few seconds the two studied each other from a distance of 150 to 200 feet, then fled in opposite directions.

Very upset, Freeman notified his bosses in Walla Walla, Washington, at once, and two hours later a group of Forest Service workers arrived at the site, which was located in Oregon near the Washington border. They found 21 footprints measuring 14 inches long and seven inches wide. They took three casts and some photographs of the prints.

On June 14 the Walla Walla station made a public statement recounting Freeman's sighting and remarking that "no determination can be made" about the identity of the creature he had claimed to see. The Forest Service also reported that it had no further plans to investigate. Still, four days later it revealed that on the 16th, Freeman and patrolman Bill Epoch had discovered about 40 new tracks in the Mill Creek Watershed on the Washington side of the border. On the 17th, Joel Hardin, a U.S. Border Patrol tracking expert and a Bigfoot doubter, examined the prints and judged them hoaxes. He said that among other suspicious features, they exhibited dermal ridges (lines in the skin), which animals do not have. He failed to mention, however, that higher primates—monkeys, apes, and human beings—do have such ridges on their toes and fingers (thus, fingerprints).

Search for Body Reveals Big Footprints

The day after Freeman's sighting, Oregon's Umatilla County Sheriff's Department sent a five-person team of volunteers to the same Tiger Creek area. The searchers were not looking for a Bigfoot, however, but for the body of a boy who had disappeared the autumn before. They went to the site because of Freeman's mention of a terrible smell, which they thought might be that of a decaying corpse. Though the team found neither the odor nor a body, it did make another discovery.

According to Art Snow, a local businessman who headed the team, the search party was able to follow the tracks beyond the 21 found by the Forest Service. In fact, Snow said, the tracks could be seen for three-quarters of a mile. The team made a cast of one of the better prints. Snow related that all evidence found by his group supported Freeman's story.

Soon afterward Washington State University anthropologist Grover Krantz, one of Bigfoot's leading scientific supporters, studied casts from both the Tiger Creek and Mill Creek Watershed areas. He also obtained a cast from Snow. After weeks of study, Krantz concluded that the prints were from "two individuals." For it seemed that one creature had a big toe larger than that of the average Bigfoot track. And the second had a "splayed-out second toe."

Aside from these differences, the prints were generally alike and typical of those connected with Bigfoot reports. The feet were about 15 inches long, and the toes were more equal in size than those of a human being. The arches were nearly flat, and there was a "double ball" at the base of the big toe.

What made the Bigfoot prints more credible was the fact that there were no human tracks around them. And the distance between the prints suggested that whatever made them had a *long* stride. In addition, Krantz noted, the tracks were so deeply pressed into the ground that most investigators felt it would take more than six hundred pounds of force to make them; no evidence of any mechanical device capable of faking this effect was found.

Other investigators, however, had serious questions about the tracks. For one thing, the prints were just too perfect. The stride did not vary, and there was no evidence of slipping up and down hillsides. When the prints were found in mud, they were not nearly as deep as they should have been if the animal weighed the 800 to 1,000 pounds that investigators estimated. (In fact, the Bigfoot prints were shallower than the tracks left by searchers' boots!) In addition, according to wildlife biologist Rodney L. Johnson, "it appeared that the foot may have been rocked from side to side to make the track." Johnson also noted that at one place where tracks were seen, "it appeared that fine forest litter ([pine] needles, etc.) had been brushed sideways from the track area in an unnatural manner."

There were also doubts about Freeman's honesty. Longtime Bigfoot tracker Bob Titmus expressed suspicions that Freeman had manufactured evidence. And—in a television interview—Freeman once admitted that he had faked Bigfoot prints!

Though the Freeman tracks may not be real, there are many other Bigfoot prints that resist explanation. John Napier wrote that in order to judge all tracks fake, we would have to believe in a conspiracy taking place "in practically every major township from San Francisco to Vancouver." If such animals *do* live in the Northwest, it seems unlikely that they can remain hidden forever!

Sources:

Beck, Fred, and R. A. Beck, *I Fought the Apemen of Mt. St. Helens,* Kelso, Washington: The Authors, 1967.

Bord, Janet, and Colin Bord, *The Bigfoot Casebook,* Harrisburg, Pennsylvania: Stackpole Books, 1982.

Green, John, *On the Track of Sasquatch,* New York: Ballantine Books, 1973.

Napier, John, *Bigfoot: The Yeti and Sasquatch in Myth and Reality,* New York: E. P. Dutton and Company, 1973.

Sanderson, Ivan T., *Abominable Snowmen: Legend Come to Life,* Philadelphia: Chilton Book Company, 1961.

Shoemaker, Michael T., "Searching for the Historical Bigfoot," *Strange Magazine,* vol. 5, 1990.

Sprague, Roderick, and Grover S. Krantz, eds., *The Scientist Looks at the Sasquatch,* Moscow, Idaho: University Press of Idaho, 1979.

MOMO

For a few days in the summer of 1972, the story of Momo, an apeman-like creature seen in and around the small Missouri town of Louisiana (population 4,600) appeared in newspapers around the country. Momo got its name from the abbreviation of Missouri—Mo.—and the first two letters of the word *monster*.

First Sighting

The Momo scare actually began in 1971. In July of that year two picnickers in a wooded area outside town reported seeing a "half-ape

and half-man" that emitted an awful smell. Stepping out of a thicket, it walked toward them while making a "little gurgling sound." They fled and locked themselves inside their car. The creature ate a peanut butter sandwich that they had left behind and strolled back into the woods. The women reported the incident to the Missouri State Police but did not come forward publicly until a year later—after many others had reported similar sightings.

Two Weeks of Terror

Momo earned its name after a series of sightings that began on July 11, 1972. That afternoon three children saw a creature "six or seven feet tall, black and hairy," standing next to a tree. It was flecked with blood, probably from the dead dog it carried under its arm. The same day a neighbor heard strange growling sounds, and a farmer found that his new dog had disappeared.

Three evenings later, the children's father, Edgar Harrison, stood talking with friends outside his home. Suddenly, they saw a "fireball" come over a nearby hill and appear to land behind an unused school-house across the street. Five minutes later a second fireball did the same thing. Not long afterward, they heard a loud growling from the hilltop, though they could not see what was making the sound. The police investigated but found nothing.

An hour or two later, as they poked around the hilltop in the darkness, Harrison and his companions came across an old building. It was befouled by the strong, rank odor linked with Momo's appearances. Later, other Louisiana witnesses reported seeing small, glowing lights, which exploded and left the same smell behind.

The scare continued for two more weeks, during which time others reported seeing a hairy biped with both ape and human features. Some even claimed that they heard voices drifting through the air. One voice said, "You boys stay out of these woods," and another asked for a cup of coffee! Footprints supposedly made by the creature were found several times, but the only one to undergo scientific examination was judged a hoax by Lawrence Curtis, director of the Oklahoma City Zoo. And a number of Louisiana residents reported seeing fireballs and other unusual objects in the air. One, described as a UFO with lighted windows, reportedly landed for five hours on a hilltop. Another was a "perfect gold cross on the moon." The family that claimed to witness this sight said it "lit up" the road "as bright as day."

That afternoon three children saw a creature "six or seven feet tall, black and hairy," standing next to a tree. It was flecked with blood, probably from the dead dog it carried under its arm.

Sources:

Clark, Jerome, and Loren Coleman, "Anthropoids, Monsters, and UFOs," *Flying Saucer Review* 19,1, January/February 1973, pp. 18-24.
Coleman, Loren, *Mysterious America,* Boston: Faber and Faber, 1983.
Crowe, Richard, "Missouri Monster," *Fate* 25,12, December 1972, pp. 58-66.

LAKE WORTH MONSTER

In the summer of 1969, residents of Forth Worth, Texas, were terrified by repeated sightings of a hairy bipedal (two-legged) monster near local Lake Worth. Though the case is not well known, it *is* one of the best documented of the many and varied **hairy biped** accounts on record. And it produced one of the very few photographs ever taken of such reported beasts.

Early on the morning of July 10, John Reichart, his wife, and two other couples showed up at a Forth Worth police station. They were so terribly frightened that—as incredible as their story sounded—the officers had no trouble believing that the six had seen something extraordinary. According to the witnesses, they had been parked along Lake Worth around midnight when a huge creature leaped out of a tree and landed on the Reicharts's car. It was, they said, covered with both scales and fur and looked like a cross between a man and a goat.

Four police cars rushed to the scene but found nothing. They were impressed, however, by the 18-inch scratch running along the side of the witnesses' car. Swearing that it had not been there before, the Reicharts were sure it was a mark left by the monster's claws.

Over the previous two months, other reports of a monster had come to police attention, but officers had dismissed the sightings as pranks. And while they believed that the Reicharts and their friends were probably victims of a prank as well, the frightening, violent nature of the event made them take the matter more seriously than they had other such encounters.

Almost 24 hours after the Reichart attack, Jack Harris was driving along on the only road going into the Lake Worth Nature Center. There he saw the creature cross in front of him. It ran up and down a bluff and was soon being observed by 30 to 40 people who had come to the area hoping to see it—lured by the *Fort Worth Star Telegram*'s headline "Fishy Man-Goat Terrifies Couples Parked at Lake Worth." Within a short time, officers from the sheriff's department were also on the scene, watching the incredible sight. But when it appeared that some of the onlookers were going to approach the creature, it threw a spare tire, rim and all, at them. Witnesses jumped back into their cars, and the beast escaped into the underbrush.

Witnesses guessed that the creature was about seven feet tall and weighed 300 pounds. It walked on two feet and had whitish-gray hair. The beast had a "pitiful cry—like something was hurting him," Harris told a reporter. "But it sure didn't sound human."

They were impressed, however, by the 18-inch scratch running along the side of the witnesses' car. Swearing that it had not been there before, the Reicharts were sure it was a mark left by the monster's claws.

Investigations

In the weeks ahead, groups of searchers, many carrying guns, made nightly trips into the woods and fields along the lake. Most who viewed the creature thought it resembled a "big white ape." It left tracks (they were not preserved) that reportedly measured 16 inches long and eight inches wide at the toes. One time searchers fired on the beast and followed a trail of blood and tracks to the edge of the water.

When it appeared that some of the onlookers were going to approach the creature, it threw a spare tire, rim and all, at them.

Another time three men claimed that it leaped on their car and got off only after the vehicle collided with a tree.

A second trio of searchers spent a week tracking the creature without ever seeing it, though they did hear its cry and smelled the awful odor associated with it. They also came upon dead sheep with broken necks—victims, they believed, of the beast. Allen Plaster, owner of a dress shop, took a fuzzy black and white photograph said to show the creature at close range.

Sightings would continue on and off for years. But the last report of the 1969 scare was made by Charles Buchanan on November 7. The man said he had been dozing inside his sleeping bag in the back of his pickup truck when something suddenly lifted him up. It was the monster! Buchanan grabbed a bag with chicken in it; the creature stuffed it into its mouth, then plunged into the lake and swam toward Greer Island.

Ultimately, Helmuth Naumer, a spokesman for the Forth Worth Museum of Science and History, and Park Ranger Harroll Rogers came to the hard-to-believe conclusion that the creature was a bobcat. Another explanation, though never confirmed, was the claim that police had caught pranksters with a costume. Now if a trickster *was* the monster, he would certainly have to have been remarkably brave or stupid, considering how many searchers were carrying guns and were ready to fire on sight.

Sources:

Coleman, Loren, *Mysterious America,* Boston: Faber and Faber, 1983.
Green, John, *Sasquatch: The Apes Among Us,* Seattle, Washington: Hancock House, 1978.

JACKO

On June 30, 1884, a strange creature was captured near the village of Yale in south-central British Columbia. It was spotted from a passing British Columbia Express train by engineer Ned Austin, who thought it was a man lying dangerously close to the tracks. He quickly brought the train to a stop. Suddenly, the "man" stood up and made a barking sound, then scrambled up one of the bluffs along the Fraser River. The train crew chased the "Indian"—which is what they thought the figure was—until they finally trapped it on a rocky ledge. Conductor R. J. Craig climbed up above the ledge and dropped a rock on the creature's head, knocking it out. The crew members then tied the beast up and brought it to town, where they put it in the jail.

According to the *Daily British Colonist* of July 4, the creature—quickly named Jacko—turned out to be "something of the gorilla type standing about 4 feet, 7 inches in height and weighing 127 pounds. He has long, black, strong hair and resembles a human being with one exception, his entire body, excepting his hands (or paws) and feet are covered with glossy hair about one inch long. His forearm is much longer than a man's forearm, and he possesses extraordinary strength." Noting that some of the local residents had reported seeing an odd creature over the past two years, the *Colonist* asked, "Who can unravel the mystery that now surrounds Jacko? Does he belong to a species hitherto unknown in this part of the country?"

In the 1950s, after **Bigfoot** reports in the Pacific Northwest captured widespread attention, newspaperman Brian McKelvie began searching for earlier press accounts of such beasts and found the Jacko story. He pointed it out to John Green and Rene Dahinden, who were just beginning what would turn out to be lifelong careers as Sasquatch hunters. He told them that this was the only surviving record of the event; other area newspapers, which might also have carried the story, had been lost in a fire. In 1958 Green interviewed an elderly Yale man, August Castle, who said that he remembered when Jacko came to town, though his parents had never taken him to the jail to view the creature.

The first book in which the Jacko story appeared was Ivan T. Sanderson's 1961 *Abominable Snowmen: Legend Come to Life*. The author found the account very convincing: "excellent ... factual ... hardly being at all speculative." From that point on, nearly every book on Sasquatch mentioned Jacko. In 1973 Dahinden and coauthor Don Hunter wrote that, according to the grandson of a man who had been a judge in Yale in 1884, Jacko "was shipped east by rail in a cage, on the way to an English sideshow." No more was heard of him, and local residents assumed that he had died along the way.

From the *Colonist*'s description, primatologist John Napier thought that Jacko sounded like "an adult chimpanzee or even a juvenile male or adult female gorilla." "But unless it was an escapee from a circus it is difficult to imagine what an African ape was doing swanning about in the middle of British Columbia," he remarked. "At that time chimpanzees were still fairly rare creatures in captivity."

The creature—quickly named Jacko—turned out to be "something of the gorilla type standing about 4 feet, 7 inches in height and weighing 127 pounds."

If Jacko was a gorilla, like the one pictured here, what was he doing in British Columbia?

During all this time, Green was still trying to get to the bottom of the mystery. He learned that microfilm of early British Columbia newspapers did exist. They were not in the British Columbia Archives, where McKelvie had looked, but they were at the University of British Columbia. And there, in a July 1884 issue of New Westminster's Mainland Guardian, he came upon these comments about Jacko from a reporter who was passing through Yale: "How the story originated, and by whom, is hard for one to conjecture. Absurdity is written on the face of it. The fact of the matter is, that no such animal was caught, and how the Colonist was duped in such a manner, and by such a story, is strange." Another newspaper, the British Columbian, reported that the story had sent some 200 people scurrying to the jail. There, it related, the "only wild man visible was [jailer] Mr. Murphy, ... who completely exhausted his patience" answering questions about the nonexistent beast.

It was these newspaper findings that finally convinced Green that Jacko had not been real. Still, there were those who were unwilling to put the story to rest. In *Pursuit* magazine, for instance, Russ Kinne argued that competing newspapers were simply trying to make the *Colonist* look bad by attacking the Jacko report as untrue.

Sources:

Bord, Janet, and Colin Bord, *The Bigfoot Casebook,* Harrisburg, Pennsylvania: Stackpole Books, 1982.

Green, John, *On the Track of the Sasquatch,* Agassiz, British Columbia: Cheam Publishing, 1968.

Halpin, Marjorie, and Michael M. Ames, eds., *Manlike Monsters on Trial: Early Records and Modern Evidence,* Vancouver, British Columbia: University of British Columbia Press, 1980.

Hunter, Don, and Rene Dahinden, *Sasquatch,* Toronto: McClelland and Stewart, 1973.

Napier, John, *Bigfoot: The Yeti and Sasquatch in Myth and Reality,* New York: E. P. Dutton and Company, 1973.

Sanderson, Ivan T., *Abominable Snowmen: Legend Come to Life,* Philadelphia: Chilton Book Company, 1961.

MINNESOTA ICEMAN

One day in the fall of 1968, a University of Minnesota zoology student named Terry Cullen phoned natural science investigator and writer Ivan T. Sanderson with an amazing story: it appeared that a **Bigfoot** body frozen in a block of ice was being displayed around the country as a carnival exhibit. Sanderson, a biologist long interested in unknown animals, was understandably excited. So was his houseguest, the famous Bernard Heuvelmans, a Belgian scientist known as the founder of the study of unexpected animals, or cryptozoology.

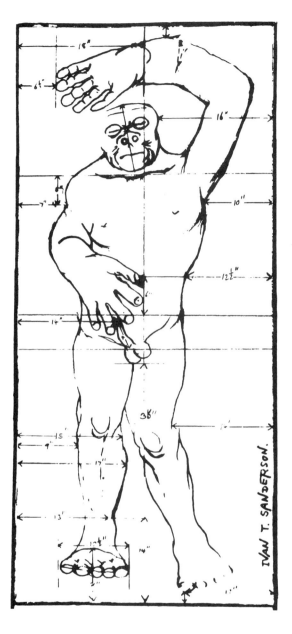

Ivan T. Sanderson's outline drawing of the Minnesota Iceman.

The two scientists lost no time in getting to tiny Rollingstone, Minnesota. There they met Frank Hansen, on whose farm the body was kept over the winter months when the carnival was not running. Hansen led them to a tiny trailer where the "Iceman"—as it was called—rested in a refrigerated coffin. Sanderson and Heuvelmans spent the next two days studying, sketching, and photographing the figure. Heuvelmans noted that it resembled an adult male human being, except that it was "entirely covered with dark brown hair three to four inches long." He also observed that the specimen showed major injuries. The left arm was twisted strangely "due to an open fracture midway between the wrist and the elbow where one can distinguish the broken ulna in a gaping wound," he reported. The creature also appeared to have been shot in the right eye. The force seemingly knocked the left eye out of its socket and blew out the back of the head.

Sanderson and Heuvelmans became convinced that the figure was what it seemed: a real body and not a model. The two scientists even examined what they thought were gas bubbles and odors escaping from the creature's slowly decaying remains.

Hansen claimed that the creature had been found floating in a 6,000-pound block of natural ice in the Russian Sea of Okhotsk. The men who discovered it were Russian seal hunters. (Though in a later version he would say they were Japanese whalers.) Eventually, according to Hansen, the body turned up in Hong Kong, where an agent of a California millionaire bought it. In time the owner rented it to Hansen, who began touring the country with it in May 1967.

Early in February 1969 Sanderson called on an old friend, John Napier, a primate expert at the Smithsonian Institution in Washington, D.C. Sharing his Iceman report and diagrams with the primatologist, Sanderson was hoping to interest the Smithsonian in joining the investigation.

As Napier would later write in his 1973 book *Bigfoot,* when he first looked at Sanderson's Iceman information, he was filled with doubt, for to him the creature seemed a crazy mixture, combining the worst traits of apes and human beings; it had "none of the best features which make these two groups extremely successful primates in their respective environments," he explained. Zoologically speaking, the Iceman did not make sense.

Still, the Smithsonian tried to obtain the specimen from Hansen. But the man said he could not provide it because its owner had, all of a sudden, reclaimed it. Hansen related that when he went back on tour it would be with a model that would "in many respects resemble" the original. After further investigation, the Smithsonian concluded that the story and the figure were a hoax.

Over the years Hansen continued to tour the United States with his Iceman exhibit, neither insisting nor denying that the figure was real. In the material he used to promote his show, however, he quoted the views of "scientists" (probably Heuvelmans and Sanderson), declaring that the Iceman was genuine.

> The creature seemed a crazy mixture, combining the worst traits of apes and human beings.

The Model Mystery

In August 1981, C. Eugene Emery, a science reporter for Rhode Island's *Providence Journal-Bulletin,* wrote an article about Hansen's exhibit, which was on display at a local shopping mall. Soon after the story appeared, Emery learned of a man named Howard Ball, then dead, who had built models for Disneyland. His specialty had been prehistoric beasts. "He made [the Iceman] here in his studio in Torrance [California]," Ball's widow, Helen, told Emery. "The man who commissioned it said he was going to encase it in ice and pass it off, I think, as a prehistoric man."

Ball's son Kenneth helped his father build the figure. He said its "skin" was half-inch-thick rubber. "We modeled it after an artist's conception of Cro-Magnon

THE CRO-MAGNON RACE

The Cro-Magnon race lived 35,000 years ago and is of the same species as modern human beings (*Homo sapiens*). Cro-Magnons stood straight and were six or more feet tall; their foreheads were high and their brains large. Discoveries of skillfully made tools, jewelry, and cave wall paintings suggest that the Cro-Magnon race had an advanced culture.

man and gave it a broken arm and a bashed-in skull with one eye popped out," he related. "As I understand it, [the man who commissioned the job] took the creature to Mexico to have the hair implanted." The Balls were pretty amused when they saw Sanderson's article in the May 1969 issue of *Argosy* and recognized their creation in the photographs published along with it.

When Emery questioned Hansen, the carnival exhibitor admitted that Ball *had* made a figure for him, but he insisted that it "was discarded." Hansen also told Emery that Ball's widow and son were mistaken about the identity of the creature shown in the *Argosy* photographs.

Had there really been a true Iceman specimen exhibited for some time before a model replaced it? Sanderson died in 1973, still convinced that the creature he and Heuvelmans had seen was not the one later shown around the country—the one they had studied had been a real animal. Similarly, Heuvelmans wrote a book defending the original Iceman (published in French and never translated into English) and even today insists that it was some kind of humanlike creature.

Sources:

Emery, C. Eugene, Jr., "News and Comment: Sasquatch-sickle: The Monster, the Model, and the Myth," *Skeptical Inquirer,* Winter 1981/1982, pp. 2-4.

Hansen, Frank, "I Killed the Ape-Man Creature of Whiteface," *Saga,* July 1970, pp. 8-11, 55-60.

Sanderson, Ivan T., "The Missing Link," *Argosy,* May 1969, pp. 23-31.

Sanderson, Ivan T., "Preliminary Description of the External Morphology of What Appeared to Be the Fresh Corpse of a Hitherto Unknown Form of Living Hominid," *Genus* 25, 1969, pp. 249-278.

Shaggy, Two-footed Creatures Abroad

- YETI

- MONO GRANDE

- ORANG-PENDEK

- ALMAS

- YEREN

- YOWIE

Shaggy, Two-footed Creatures Abroad

YETI

Southern Asia's Himalayan Mountains extend in a 1,500-mile arch across northern India, Nepal, Sikkim, Bhutan, and the southern end of Tibet. In these mountains, according to many reports, lives the legendary yeti. The first English printed reference to a strange bipedal (two-legged) creature that roamed the Himalayas may have been in the *Journal of the Asiatic Society of Bengal* in 1832. There B. H. Hodgson, British Resident of the Court of Nepal, recalled an experience he had while collecting specimens in a northern Nepal province. His native guides came upon an erect, tailless creature with long, dark hair all over its body. Thinking it was a demon, they fled in terror. Hodgson, however, thought it might be an orangutan.

In 1889 Major L. A. Waddell became the first Westerner (non-Asian) to discover a mysterious humanlike footprint in the Himalayan snows. His Sherpa (Tibetan) guides told him that the track, found at 17,000 feet, was made by a hairy wild man that lived in the area. While noting in *Among the Himalayas* (1899) that "the belief in these creatures is universal among Tibetans," Waddell complained that those who told him the tales could never present any evidence or an eyewitness. The major himself believed that such wild men were really "great yellow snowbears."

Lieutenant Colonel C. K. Howard-Bury, who led a fact-finding expedition up Mount Everest in September 1921, came upon a large number of such mysterious prints. Three times the size of human tracks, the footprints were found at 20,000 feet on the side of the mountain that faces northern Tibet. In his official report, Howard-Bury wrote down

Footprints of the abominable snowman?

the Sherpa word for the creature, which meant "manlike thing that is not a man." But he transcribed it incorrectly, and a *Calcutta Statesman* writer mistranslated the word as "abominable snowman." The "abominable snowman" captured the English-speaking public's imagination due to widespread newspaper coverage of C. K. Howard-Bury's expedition report. Howard-Bury, however, believed the tracks "were probably caused by a large 'loping' grey wolf, which in the soft snow formed double tracks rather like those of a barefooted man."

Four years later N. A. Tombazi, a British photographer and member of the Royal Geographical Society, saw a strange creature in the

Himalaya range. The sighting took place near the Zemu Glacier, at 15,000 feet. He reported: "The intense glare and brightness of the snow prevented me from seeing anything for the first few seconds; but I soon spotted the 'object' referred to, about two to three hundred yards away down the valley to the east of our camp. Unquestionably, the figure in outline was exactly like a human being, walking upright and stopping occasionally to uproot or pull at some dwarf rhododendron bushes. It showed up dark against the snow and, as far as I could make out, wore no clothes. Within the next minute or so it had moved into some thick scrub and was lost to view."

Two hours later, as Tombazi's group returned to its camp, he went to check the area where he had seen the creature. There he found 16 footprints "similar in shape to those of a man," though smaller in size. Still, the photographer had no doubt that they belonged to a creature that walked on two legs.

From these accounts—plus more detailed ones from native witnesses—the yeti (from the Sherpa word *yeh-teh,* meaning "that thing") stirred worldwide interest. Since then it has been the subject of countless expeditions, arguments, and theories. Evidence has been sketchy, and it does not appear that an answer to the yeti mystery will be found anytime soon.

Ivan T. Sanderson's 1970 drawing of the yeti.

Probably the most interesting sighting by a Westerner took place on Mount Annapurna in 1970. The witness, well-known British mountaineer Don Whillans, was looking for a campsite one evening when he heard odd cries. His Sherpa companion said that they were a yeti's call, and Whillans caught a glimpse of a dark figure on a distant ridge. The next day he found humanlike tracks sunk 18 inches into the snow. That night, sensing the creature's presence, he looked out of his tent and saw in the moonlight an ape-shaped animal as it plucked at tree branches. He watched it for 20 minutes through binoculars before it wandered away.

Two Types of Yetis

The Sherpas (Tibetans) have described two types of yetis: the *dzu-teh* ("big thing"), seven to eight feet tall, and the *meh-teh,* which ranges between five and six feet. Sightings of the meh-teh are reported far more often than those of the larger creature, and most people think of the meh-teh as the "abominable snowman."

Zoologist Edward W. Cronin, Jr., gave this general description of the meh-teh: "Its body is stocky, apelike in shape, with a distinctly human quality to it.... It ... is covered with short, coarse hair, reddish-brown to black in color, sometimes with white patches on the chest. The hair is longest on the shoulders. The ... teeth are quite large ... and the mouth is wide. The shape of the head is conical, with a pointed crown. The arms are long, reaching almost to the knees. The shoulders are heavy and hunched. There is no tail."

Physical Evidence?

Most scientific and serious examinations of the yeti question focus on the footprints. For no matter how poor a sighting is or how questionable a witness may be, the tracks—without a doubt—exist. Some researchers explain them away as the prints of known animals like snow leopards, foxes, or bears, which have become distorted into "yeti" shapes through melting. This has become a popular theory. Yet primate scientist John Napier, no great yeti supporter himself, wrote that "there is no real experimental basis for the belief that single footprints can become enlarged and still retain their shapes, or that discrete prints can run (or melt) together to form single large tracks." In any case some fresh tracks were found by Cronin and physician Howard Emery during a 1972 expedition in far-eastern Nepal. The tracks were found before the sun, wind, and weather had a chance to alter them.

Still, other types of yeti evidence have proved disappointing. Members of a 1954 *London Daily Mail* expedition, for instance, examined a "yeti scalp" said to be 350 years old, preserved as a kind of sacred object in a Tibetan lamasery (a house of Lamaist monks). Four years later the members of an expedition led by Texas oilman Tom Slick also looked at it and another specimen. And in 1960, in a much-followed expedition sponsored by the publishers of the *World Book Encyclopedia,* Sir Edmund Hillary (the first to scale the Himalayan peak Mt. Everest, the world's tallest mountain, and a famous yeti nonbeliever) was able to obtain a third specimen. After a complete examination, most scientists agreed that it was the skin of a goat antelope.

The next day he found humanlike tracks sunk 18 inches into the snow. That night, sensing the creature's presence, he looked out of his tent and saw in the moonlight an ape-shaped animal as it plucked at tree branches.

Tom Slick, a wealthy Texas oilman, led expeditions that collected yeti evidence, including what may be parts of the hand of an unknown biped.

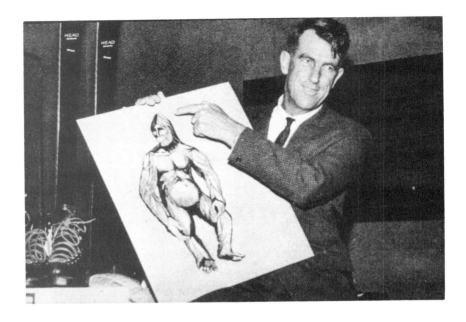

Sir Edmund Hillary (the first to scale the Himalayan peak Mt. Everest, the world's tallest mountain) cast doubt on the existence of yetis.

The 1958 Slick expedition also resulted in two specimens of what were supposed to be yeti hands. One proved to be the paw and forearm of a snow leopard. But the other specimen appeared to be something else again. It may be the single best piece of evidence for the yeti's existence.

In early 1959 expedition member Peter Byrne was allowed into a lamasery at Pangboche, Nepal, where he had learned that the bones of a hand supposedly belonging to a yeti were kept. The monks insisted that the hand not leave the building, but Byrne had other ideas. In a carefully thought-out plan, Byrne got the lamas to let him examine the specimen by himself. He had with him parts of a human hand; he secretly switched its thumb and part of the index finger with those of the Pangboche specimen. When Byrne left the monastery, the lamas had no reason to suspect that the yeti hand had been partially dismantled and rebuilt.

The stolen samples, which also included a piece of skin, had to be smuggled out of the country. Crossing the Nepal border was managed easily but getting the contraband out of India, where customs officials were far stricter, looked difficult. But as luck would have it, two close friends of expedition cosponsor Kirk Johnson were staying at a Calcutta hotel. Hollywood film actor Jimmy Stewart and his wife, Gloria, wrapped the samples in underwear, buried them deep in their luggage, and took them to London undetected. There they delivered them to Johnson.

Shaggy, Two-footed Creatures Abroad

British primatologist W. C. Osman Hill, who had supplied Byrne with the human hand parts, received the specimen on February 20. His first impression was that it was "human" and he was disappointed. Later, however, he thought that the thumb and phalanx (finger bone) appeared less than fully human—that, as unlikely as it seemed, they might even be the remains of a Neanderthal man. Two other scientists who examined the samples also admitted that they were puzzled. Zoologist Charles A. Leone regretted that he could not "make a positive identification." And anthropologist George Agogino told writer Gardner Soule, "Many people who have examined this hand feel that it is a human hand with very primitive characteristics.... I do not feel that this hand is a normal human hand at all.... It is highly characteristic, however, of all the giant anthropoids [large, tailless, semi-upright apes]." Blood tests of the skin sample showed it was from no known primate or human.

Because of the unscrupulous way in which the samples were collected, they received no news coverage or publicity. Thus when Hillary made his expedition in 1960, he was not aware that the Pangboche hand had been tampered with. In his opinion, the yeti was something of a joke, and he declared with much amusement that the Pangboche hand was "essentially a human hand, strung together with wire, with the possible inclusion of several animal bones." This, of course, is exactly what it was after Bryne got through with it. Had Hillary looked more closely at the "animal" bones—instead of the added human ones—he might have been as puzzled by them as Hill and the other scientists.

The negative results of Hillary's expedition quieted scientific and public interest in the "abominable snowman," though a few books, magazine articles, and trips into the Himalayas (such as Cronin's in 1972-74) would revive the subject from time to time. In a comical case in 1986, an English traveler took what he truly believed was a photograph of a yeti—but later investigation showed that the "yeti" was a mountain rock. In February and March of the same year, the New World Explorers Society collected reports of recent sightings by Himalayan residents and returned with samples of what was supposed to be yeti hair: "long, black, and coarse." The samples were turned over to the International Society of Cryptozoology for study, but the results have yet to be released.

Fecal droppings supposedly left by yetis are another kind of evidence in the abominable snowman mystery. Samples collected by Slick's expedition contained eggs of an unknown worm parasite.

An English traveler took what he believed was a photograph of a yeti—but later investigation showed that the "yeti" was a mountain rock.

Famous cryptozoologist Bernard Heuvelmans remarked, "Since each species of mammal has its own parasites, this indicated that the host animal is also equally an unknown animal."

Those who study the yeti believe that the dzu-teh, the larger version of the beast, is probably a blue bear. If there truly is a yeti, it is almost certainly the meh-teh. And those who believe the yeti is real agree that it does not live in the high mountain snows, but in the mountain forests below the snowfields.

With this in mind, Nicholas Warren thinks that yeti sightings can be credibly explained: the animal seen is really "a vegetarian ape, occasionally straying from the forests into the high snowfields." Yet yeti experts seem to favor a more extraordinary explanation: they believe that it (as well as other reported but unrecognized apeman creatures like Bigfoot and the Chinese wild man) is a surviving *Gigantopithecus.* This was a large prehistoric ape the fossil remains of which have been uncovered in, among other places, the Himalayan foothills.

Shaggy, Two-footed Creatures Abroad

Slick reported that when he showed native witnesses photographs of different animals and asked which one the yeti most resembled, the choices were always the same. The natives first chose "a gorilla standing up." The next selection was "an artist's drawing of a prehistoric apeman, *Australopithecus.*" And their third pick was "an orangutan standing up, which they liked particularly for the long hair."

Sources:

Coleman, Loren, *Tom Slick and the Search for the yeti,* Boston: Faber and Faber, 1989.

Cronin, Edward W., Jr., "The yeti," *Atlantic Monthly,* November 1975, pp. 47-53.

Heuvelmans, Bernard, *On the Track of Unknown Animals,* New York: Hill and Wang, 1958.

Hillary, Sir Edmund, "Epitaph to the Elusive Abominable Snowman," *Life,* January 13, 1961, pp. 72-74.

Napier, John, *Bigfoot: The yeti and Sasquatch in Myth and Reality,* New York: E. P. Dutton and Company, 1973.

Sanderson, Ivan T., *Abominable Snowmen: Legend Come to Life,* Philadelphia: Chilton Books, 1961.

Soule, Gardner, *Trail of the Abominable Snowman,* New York: G. P. Putnam's Sons, 1966.

MONO GRANDE

The mystery of the *mono grande*—Spanish for "big monkey"—has puzzled zoologists for years. The only known primates in North and South America are small, long-tailed monkeys. Yet sightings of larger, tailless, anthropoid (manlike) apes have been reported from time to time in the remote regions of the northern climates of South America. In fact, a photograph of such a creature is at the heart of the argument that scientists have waged over the subject.

The Fateful Expedition

Between 1917 and 1920 an expedition led by Swiss oil geologist Francois de Loys explored the swamps, rivers, and mountains west and southwest of Lake Maracaibo near the Colombia-Venezuela border. The explorers suffered great hardship, and a number of them died from disease or were murdered by hostile natives. In the last year, what remained of the expedition was camped on the banks of a branch of the Tarra River. Suddenly, two creatures, apparently male and female, stepped out of the jungle. De Loys at first thought they were bears, but

Unknown ape
photographed by
Francois de Loys in
Venezuela/Colom-
bia area of South
America.

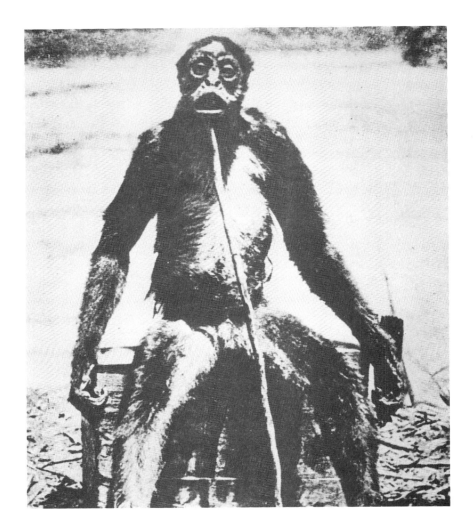

as they neared the camp, he could see that they were apes of some sort, about five feet in height. His account does not, however, include the important detail of whether each walked on two or four feet.

The creatures, which appeared to be very angry, broke off branches from nearby trees and swung them as weapons, meanwhile crying out and moving wildly. Finally, they emptied their bowels into their hands and hurled the results at the party, who by now had their rifles to their shoulders. In the gunfire that followed, the female was killed, and the wounded male escaped back into the underbrush.

Though no one in the expedition was a zoologist, everyone somehow understood that the animal was extraordinary. Even the native guides swore they had never seen anything like it. Propping it up with a stick, members sat the creature on a gasoline crate and took a picture of it from ten feet away. According to de Loys, "Its skin was afterward removed, and its skull and jaw were cleaned and preserved." Though de Loys did not state it explicitly, it is believed that he and his starving companions ate the animal's flesh! Later, the creature's remains were reportedly lost. Of the 20 members who began the expedition, only four survived.

Photo Draws Scientist's Attention

The photograph, however, was discovered by a friend of de Loys—anthropologist George Montandon—when he was looking through the geologist's records and other expedition materials. Montandon considered the picture so important that he immediately started planning a trip to "the area in question to find the great ape of America." De Loys, he noted, showed no interest in publishing the photograph or talking publicly about it. Only when Montandon insisted was it brought to the world's attention, when he reported it in three French scientific journals in 1929. In his articles Montandon honored de Loys by offering the formal name *Ameranthropoides loysi* for what he believed was a new animal. That same year de Loys told his story publicly for the first time in the popular magazine *Illustrated London News* (June 15).

The Doubter's View

Hardly any time had passed before doubters attacked de Loys's photograph and Montandon's judgment in bringing it before the public. Leading the pack was renowned British anthropologist Sir Arthur Keith, who felt that de Loys had taken a picture of a smaller, tailless animal, the spider monkey—which did live in South America—and tried to pass it off as something more interesting. Keith wrote: "A photograph of the animal from behind would have clinched matters, but the only photograph taken was one of the front—the animal being placed in a sitting position on a box of unknown size and with no standard object in or near the body of the animal to give a clue to the dimensions of its parts."

But if the gasoline crate on which the ape sat was a common one (and there is no reason to think that it was not), then the size of the creature could be guessed. Such crates measure a standard 20 inches—making the figure atop it about 5'2" or 5'3" tall! That would be a

highly irregular spider monkey, for the largest known specimen measured 3'7".

Keith also sneered at de Loys's claim that the animals had thrown their feces at the party, as if the idea were ridiculous. But spider monkeys and some apes do such things when faced with enemies. Keith also found it suspicious that de Loys had lost all evidence but the photograph. Yet considering the expedition's problems, it is hardly surprising that hanging on to the creature's remains was not of paramount importance; de Loys and his men were more concerned with staying alive.

Indeed, nothing about the case suggested that de Loys was trying to pull off a hoax; it was only Montandon's interest that kept the photo and its story from being buried forever. Still, the arguments of Keith, however refutable, remained the last word on the subject. In 1951, for example, *Natural History* reported that Keith had "easily demolished the 'new anthropoid.'"

The Nature of the Beast

Michael T. Shoemaker, a researcher of strange events, examined de Loys's photograph at length and published his observations in *Strange Magazine* in 1991. He wrote that the creature in the picture had features of both spider monkeys and anthropoid apes. Its flat nose, ridged eye sockets, long hair, and very long fingers and toes were monkeylike, but its body resembled a gibbon's, and its limbs and small thumbs looked like those of an orangutan. Most important to Shoemaker was the shape of the animal's head: its jaws were much heavier and more powerful than a spider monkey's, and its forehead was far more developed than that of any monkey type he knew.

De Loys may have taken the only photograph of a possibly unknown species of anthropoid ape, but he was not the only one to report seeing such an animal. The first printed report appears in the 1553 records of Pedro de Cieza de Leon, who mentions native sightings and refers to a Spaniard who said "he had seen one of these monsters dead in the woods." In *An Essay on the Natural History of Guiana* (1769), Edward Bancroft relates Indian reports of creatures "near five feet in height, maintaining an erect position, and having a human form, thinly covered with short, black hair." In 1860's *The Romance of Natural History,* Philip Gosse wrote that he felt fairly sure that a "large anthropoid ape, not yet recognized by zoologists," did exist in South America.

In 1876 Charles Barrington Brown, explorer of what was then British Guiana (now Guyana), wrote about a creature that natives called the *Didi,* a "powerful wild man, whose body is covered with hair, and who lives in the forest." He heard it spoken of many times and even saw its footprints once, or so they were identified to him. Sighting reports of anthropoid apes have continued into the twentieth century.

In 1968 explorer Pino Turolla, while traveling in the jungle-covered mountains of eastern Venezuela, was told of the mono grande. His guide said that three of the creatures, using branches for clubs, had attacked him and killed his son several years earlier. On his return to the United States, Turolla researched the matter and came upon the de Loys photograph; on his next expedition that year he showed the picture to his guide, who said that, yes, this was what the mono grande looked like. Turolla had the guide take him to the canyon where the awful attack had taken place—and there, after hearing eerie howling sounds, the explorer saw a pair of apelike creatures about five feet tall running upright on two legs. Turolla claimed a second, briefer sighting two years later while on an expedition in the Andes Mountains of Ecuador.

Turolla had the guide take him to the canyon where the awful attack had taken place—and there, after hearing eerie howling sounds, the explorer saw a pair of apelike creatures about five feet tall running upright on two legs.

The most recent published sighting is from Guyana in 1987. The witness, plant scientist Gary Samuels, was gathering fungi from the forest floor when he heard footsteps. Expecting to see a Guyanese forest worker, he was startled to look up and find a five-foot-tall, upright ape "bellowing an occasional 'hoo' sound."

Sources:

Camera, I., and G. H. H. Tate, "Letters: The 'Ape' That Wasn't an Ape," *Natural History* 60,6, June 1951, p. 289.

Picasso, Fabio, "More on the Mono Grande Mystery," *Strange Magazine* 9, spring/summer 1992, pp. 41, 53.

Shoemaker, Michael T., "The Mystery of the Mono Grande," *Strange Magazine* 7, April 1991, pp. 2-5, 56-60.

Turolla, Pino, *Beyond the Andes,* New York: Harper and Row, 1980.

ORANG-PENDEK

Sumatra is a large Indonesian island graced by millions of acres of rain forest. It is the home of the gibbon (a small, tailless ape), the orangutan, and the sun bear (the last of a species of bear that stands on its hind feet, though it does not run on them). And many islanders also speak of another extraordinary animal in these forests: the *orang-pendek,* or "little man." (Some call it the *sedapa.*)

Orang-pendeks are said to stand between two and a half and five feet tall and to be covered with short dark hair, with a thick, bushy mane that goes halfway or farther down the back. Their arms are shorter than an ape's, and—unlike Sumatra's other apes—they more often walk on the ground than climb in trees. An orang-pendek's footprint is like that of a small human being, only wider. It eats fruits and small animals.

Early Sightings

Witnesses often mention how much orang-pendeks look like human beings. A Dutch settler named Van Herwaarden said he came upon one in October 1923; and though he had his rifle and was an

"Its eyebrows were frankly moving; they were of the darkest color, very lively, and like human eyes."

experienced hunter, he reported, "I did not pull the trigger. I suddenly felt that I was going to commit murder." He observed that "the creature's brown face was almost hairless, whilst its forehead seemed to be high rather than low. Its eyebrows were frankly moving; they were of the darkest color, very lively, and like human eyes. The nose was broad with fairly large nostrils, but in no way clumsy.... Its lips were quite ordinary, but the width of its mouth was strikingly wide when open. Its canines [pointed teeth] showed clearly from time to time.... They seemed fairly large to me, ... more developed than a man's.... I was able to see its right ear which was exactly like a little human ear. Its hands were slightly hairy on the back." Van Herwaarden believed that the creature he had come upon was a female and that she was about five feet tall.

Because primatologists (scientists who study primates) have never been shown a living or dead specimen, most have rejected such eyewitness reports as tricks or mistaken identifications of orangutans or gibbons. Some footprints thought to be those of orang-pendeks were later identified as belonging to sun bears, for example.

New Investigations

In the summer of 1989, British travel writer Deborah Martyr visited the rain forests of southwestern Sumatra. There her guide told her about orang-pendeks and where they could be found. When Martyr expressed her doubts about such a creature, the guide related his own two sightings.

Fascinated, Martyr began to interview residents in the area and collected many sighting reports. "All reports included the information that the animal has a large and prominent belly—something not mentioned in previous literature on the subject," she wrote. Some reported that the mane was dark yellow or tan, others that it was black or dark gray. When Martyr suggested to witnesses that the creatures they saw might really be orangutans, gibbons, or sun bears, they strongly denied it.

Martyr herself traveled to the south edge of the Mount Kerinci region, where she was told the creatures were often seen. Though she did not have a sighting of her own, she did find tracks. Of one set she noted, "Each print was clearly delineated, the big toe and four smaller toes easily visible. The big toe was placed as it would be in a human foot." Each measured about six inches in length and four inches wide at the ball of the foot. Martyr added that "if we had been reasonably close to a village, I might have momentarily thought the prints to be those of a healthy seven-year-old child. The ball of the foot was, however, too broad even for a people who habitually wear no shoes."

Because of falling rain and poor lighting, the photographs Martyr took of the tracks did not turn out well. But she did manage to make a plaster cast of a footprint, which she took to the headquarters of the Kerinci Seblat National Park in Sungeipenuh. The park's director had ignored earlier orang-pendek reports because, as he told Martyr, the local people were "simple." But when he and his workers saw the cast, they agreed that it was of an animal they did not know.

Unfortunately, the cast—a promising piece of evidence—was sent to the Indonesian National Parks Department, where it was never seen or heard about again. Martyr repeatedly tried to get a judgment on it or at least have it returned to her, but she had no luck at all.

Martyr hopes to continue her investigations. She is 80 percent sure that the orang-pendek exists in the high rain forests of southwestern Sumatra. "If it is ground-dwelling and elusive," she says, "this could explain how it has escaped zoological notice, and is known only to the native people."

Sources:

Heuvelmans, Bernard, *On the Track of Unknown Animals,* New York: Hill and Wang, 1958.
Martyr, Deborah, "An Investigation of the *Orang-Pendek,* the 'Short Man' of Sumatra," *Cryptozoology* 9, 1990, pp. 57-65.

ALMAS

In the Mongolian language, *Almas* means "wild man." These strange creatures, half human, half ape, reportedly live in the Altai Mountains in western Mongolia and in the Tien Shan of nearby Sinkiang in the People's Republic of China.

History

The earliest known printed account of Almases appeared in a journal written by Bavarian nobleman Hans Schiltberger. In the 1420s Schiltberger traveled through the Tien Shan range as a prisoner of the Mongols. "In the mountains themselves live wild people, who have nothing in common with other human beings," he recorded. "A pelt covers the entire body of these creatures. Only the hands and face are free of hair. They run around in the hills like animals and eat foliage [leaves] and grass and whatever else they can find." Schiltberger saw two of them for himself, a male and a female that a local warlord had caught and given as gifts to the Bavarian's captors.

A late-eighteenth-century Mongolian manuscript on natural history contains a drawing of an Almas. The figure is identified as a "man-animal." All of the other illustrations in the book are of real animals, demonstrating that Almases were not considered legendary beings; they were viewed as ordinary creatures of flesh and blood.

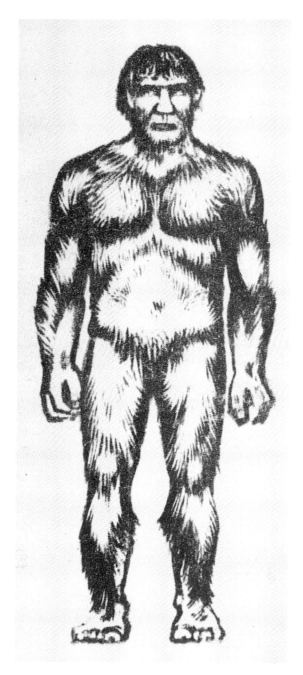

"Wild man" (Almas) seen in Buinaksk, Dagestan, in the former U.S.S.R. in 1941

Professor Tsyben Zhamtsarano conducted the first real scientific study of Almases, collecting reports from nomads and others in the remote areas where the creatures—adults and children—were said to live. He plotted the sightings on maps and brought an artist with him on his field trips to interview witnesses. But while living in Leningrad, Russia, in the 1930s under Soviet dictator Joseph Stalin, Zhamtsarano was imprisoned for his interest in Mongolian folklore, and he died captive in 1940. The records of his Almas research have been lost.

Still, one of Zhamtsarano's associates, Dordji Meiren, said that their research showed fewer Almas sightings in the later decades of the nineteenth century. The Almases had largely disappeared from Inner Mongolia and southern Outer Mongolia; it appeared that they were migrating westward to escape ever-growing civilization. Another early researcher, anatomy expert V. A. Khakhlov, shared his Almas findings with the Russian Imperial Academy of Sciences in 1913. These studies, too, no longer exist.

In the 1920s writer M. K. Rosenfeld heard about Almases during a trip across Mongolia. He later used the creatures in the plot of his adventure novel, *The Ravine of the Almases,* published in 1936. By that time another Mongolian scholar, Y. Rinchen, was conducting his own Almas research, and in the 1950s—with renewed interest in the Himalayan **yeti** or "abominable snowman"—the Soviet Academy of Sciences set up a Commission for the Study of the Snowman Question. The commission's leader, Boris Porschnev, encouraged Rinchen to publish some of his findings. Like researchers before him, Rinchen concluded that the Almas population was shrink-

Shaggy, Two-footed Creatures Abroad

ing and retreating. Since then other Russian and Mongolian scholars have published Almas studies.

Zhamtsarano associate Meiren claimed to have seen an Almas skin being used as a ceremonial carpet in a Buddhist monastery in the southern Gobi region of Mongolia. The creature had been skinned by a straight cut down the spine, so its features were preserved. The body had red, curly hair, and there was long hair on the head, but the face was hairless except for eyebrows. The nails at the ends of the toes and fingers were humanlike in appearance.

A Typical Almas

Adult Almases have been described as five feet or slightly taller, hairy, with noticeable eyebrow ridges, a receding chin, a jaw that juts out—and a shy manner. They eat small mammals and wild plants and use simple tools but have no language. According to British anthropologist Myra Shackley, their "very simple lifestyle and the nature of their

Some scientists believe Almases are surviving forms of prehistoric human beings, such as the Neanderthal man pictured here.

appearance suggests strongly that Almas[es] might represent the survival of a prehistoric way of life, and perhaps even of an earlier form of man. The best candidate is undoubtedly Neanderthal man." Neanderthals lived between 40,000 and 100,000 years ago. The classic Neanderthal had a large, thick skull with heavy brow ridges, a sloping forehead, and a chinless jaw. The link between Neanderthal man and human beings is unclear.

Chris Stringer is another British anthropologist open to the idea of the Almases' existence. Still, he finds many things in Almas reports that do not fit "accepted ideas about the Neanderthals." Among them are the creatures' physical features, like "bent knees, ... turned-in feet, ... long arms, forearms, hands and fingers, [and] small flat noses," as well as their "lack of language, culture, meat-eating and fire."

Sources:

Bord, Janet, and Colin Bord, *The Evidence for Bigfoot and Other Man-Beasts,* Wellingborough, Northamptonshire, England: The Aquarian Press, 1984.
Sanderson, Ivan T., *Abominable Snowmen: Legend Come to Life,* Philadelphia: Chilton Book Company, 1961.
Shackley, Myra, *Still Living?: Yeti, Sasquatch and the Neanderthal Enigma,* New York: Thames and London, 1983.
Stringer, Chris, "Wanted: One Wildman, Dead or Alive," *New Scientist,* August 11, 1983, p. 422.

YEREN

For centuries, in remote areas of central and southern China, residents and travelers have spoken of something called a "wild man," or *yeren.* In early writings, the creature was also referred to as a "hill ghost," "mountain monster," "man bear," or something "monkeylike, but not a monkey." One seventeenth-century account from the Hubei province offered this description: "In the remote mountains of Fangxian County, there are rock caves, in which live hairy men as tall as three meters [over nine feet]. They often come down to hunt dogs and chickens in the villages. They fight with whoever resists."

Although in the late 1950s some Chinese scientists took an active interest in the **yeti**—the "abominable snowman" of the Himalayas—they gave little notice to their own hairy giants, for they considered

yeren merely popular folklore figures. Those who claimed contact with the creatures were mostly peasants or soldiers in the provinces (though two scientists also claimed to have seen the animal).

Sightings

Biologist Wang Tselin reported seeing a yeren killed in 1940 in the Gansu area. He described the creature as a female about six and a half feet tall and covered with grayish-brown hair, the human and ape features of her face reminding him of a type of prehistoric human being. In 1950 geologist Fan Jingquan said he twice observed two yeren, seemingly a mother and son, in a mountain forest.

The first official examination of the yeren question took place in 1961, after the reported killing of a female by road builders in a thick forest in the Xishuang Banna area. By the time members of the Chinese Academy of Sciences got to the site, though, the body was no longer there, and the scientists concluded that the animal was nothing more than a gibbon (a small tailless ape). This verdict halted government interest in the yeren for the next 15 years. Two decades later, Zhou Guoxing, an anthropologist with the Beijing Natural History Museum, interviewed a journalist who had been connected with the 1961 investigation. "He stated that the animal which had been killed was not a gibbon, but an unknown animal of human shape," Zhou related.

A 1976 incident involving a yeren and several witnesses renewed interest in the creature and brought it worldwide attention for the first time. Early on the morning of May 14, six local government officials driving home from a meeting spotted a "strange, tailless creature with reddish fur" on a country highway near Chunshuya, Hubei province. Switching the headlights to high, the driver followed the animal as it tried to escape up a slope along the roadside. But it slipped and landed right in front of the jeep. The five passengers jumped out and surrounded the beast, which was now on all fours and staring directly into the lights.

Afraid to get too close, the unarmed witnesses moved to within six feet of the creature. One of them tossed a rock at its rear end, causing it to stand briefly. This frightened the group, which retreated. The animal then lumbered away, this time making a successful climb up the slope.

Group members described the creature as more than six feet tall. Covered with thick brown-red and purple-red wavy hair, it had a fat belly and large buttocks. Its eyes looked almost human. Still it had the large ears of an ape and the jutting snout of a monkey.

Investigations Commence

The Chunshuya sighting caused a stir at the Chinese Academy of Sciences, which sent 110 investigators into the field the next year. They

Switching the headlights to high, the driver followed the animal as it tried to escape up a slope along the roadside. But it slipped and landed right in front of the jeep.

Shaggy, Two-footed Creatures Abroad

focused their efforts on the forests of Fang County and the Sennongjia area of Hubei—a huge forest preserve of steep mountains and deep valleys where a wide variety of rare and exotic animals, including the giant panda (discovered only in 1869), lived. None of the investigators had a personal sighting, but they interviewed witnesses and collected footprints, hair, and feces that supposedly originated with yeren.

Zhou Guoxing, one of the expedition leaders, later noted that there seemed to be two types of yeren. There was "a larger one of about two meters in height [around six feet] and a smaller one, about one meter in height [three feet]." The two types also had a very different footprint, with the large yeren's appearing "remarkably similar to that of a man" and the smaller creature's being "more similar to the footprint of an ape or monkey, with the largest toe evidently pointing outwards."

The Smaller Yeren

To Zhou and others, the existence of the smaller type of yeren seemed almost certain. The anthropologist pointed out that both living and dead specimens of the creature might already be in scientists' hands. One was killed on May 23, 1957, near the village of Zhuanxian in Zhejiang province. A biology teacher there preserved its hands and feet. When Zhou learned of this in 1981, he went and collected the specimens. After a long study he concluded that they "belonged to a kind of large stump-tailed monkey unknown to science." He thought the creature was a stump-tailed macaque. Not long afterward just such an animal was captured in the Huang Mountain region and taken to the Hefei Zoo.

On the other hand, Ohio State University anthropologist Frank E. Poirier believed that yeren reports were probably sightings of a rare, endangered animal that lived in the region but was seldom seen: the golden monkey. But after a 1989 expedition, he changed his mind. Noting that local Chinese residents used the term "yeren" (which means wild man) to describe a number of animals—including bears, apes, and monkeys—he felt that an unknown yeren might exist after all. Poirier himself, in fact, was once mistaken for a yeren after villagers who had never seen a Caucasian man discovered him, half-clothed, napping by a river. Poirier and his colleague J. Richard Greenwell also thought that the smaller yeren might be "orangutans, ... either the known species, or more likely, a related species—perhaps even a fossil form—populations of which may [have survived] in rugged and isolated pockets of the country."

The Larger Yeren: Bigfoot's Cousin?

While the smaller yeren is of major interest to primate scientists, the other yeren seems to be a find of a different nature: a Chinese cousin of North America's **Bigfoot**. Walking on two legs, it stands between six and eight feet tall and has a strikingly humanlike face. One witness gave this description of a male yeren to Chinese Academy of Sciences researchers: "He was about seven feet tall, with shoulders wider than a man's, a sloping forehead, deep-set eyes and ... his jaw jutted out.... His hair was dark brown, more than a foot long and hung loosely over his shoulders. His whole face, except for the nose and ears, was covered with short hairs. His arms hung below his knees.... He didn't have a tail, and the hair on his body was short."

Investigators have collected dozens of hairs, supposedly from yeren, and examined them in laboratories. Studying samples from different areas of China, physicists at Fudan University found that they were indeed different—with the proportion of iron to zinc 50 times greater than that found in human hair and seven times greater than in the hair of known primates. According to Poirier and Greenwell, this seems to suggest that "some specific Wildman hairs derive from a higher primate not yet known to zoology." A second testing of the samples by others brought the same results.

Biologists at East China Normal University used a scanning electron microscope to examine yeren hairs, which they compared to those of humans and primates. They concluded that the yeren hairs were different from both the human and primate samples, but they most closely matched the human hair. This is not surprising given that eyewitness accounts of the creature describe it as more human than ape.

Zoologists receptive to the idea that a large yeren may exist have a favorite theory (shared by Bigfoot researchers). They suggest that such a creature is a surviving *Gigantopithecus,* a giant early primate that is thought to have walked upright, on two legs. Believed to have become extinct in China some 300,000 years ago, the animal had existed for about eight million years. "It takes only a little 'push' to propose its survival another half-million years to the present time," Poirier and Greenwell wrote, pointing out that the giant panda—which is just as old—shares the same native home.

Sources:

Bord, Janet, and Colin Bord, *The Evidence for Bigfoot and Other Man-Beasts,* Wellingborough, Northamptonshire, England: The Aquarian Press, 1984.

Coleman, Loren, *The Yeti and the Yeren,* Portland, Maine: The Author, 1992.

Greenwell, J. Richard, and Frank E. Poirier, "Further Investigations into the Reported *Yeren,*—The Wildman of China," *Cryptozoology* 8, 1989, pp. 47-57.

Poirier, Frank E., Hu Hongxing, and Chung-Min Chen, "The Evidence for Wildman in Hubei Province, People's Republic of China," *Cryptozoology* 2, winter 1983, pp. 25-39.

YOWIE

The yowie is Australia's version of **Bigfoot.** The existence of *this* ape-man-like creature would be more incredible than most others reported around the world, for Australia has been separated from the Asian continent for some 70 million years. This disconnection of land masses occurred far too long ago for anthropoid apes (those that are large, tailless, and semi-upright) to have crossed over and evolved into the kind of creature Australians have reported seeing for years.

The Yahoo

During the nineteenth century European immigrants to Australia noted that the native aborigines were terrified of something called a *yahoo* or *devil-devil.* The writer of an article in an 1842 issue of *Australian and New Zealand Monthly Magazine* wondered if the creatures were merely imaginary, though some Australian naturalists believed that yahoos were real animals. Because of the creatures' "scarceness, sly-ness, and solitary habits," according to the article, "man has not succeeded in obtaining a specimen." It concluded that yowies were "most likely to be one of the monkey tribe." Two years later, in her *Notes and Sketches of New South Wales During Residence in the Colony from 1839 to 1844,* Mrs. Charles Meredith noted the aborigines' fear of the yahoo, and reported that the creature "lives in the tops of the steepest and rockiest mountains, which are totally inaccessible to all human beings."

After awhile, Australian settlers began seeing the yahoo, too. An 1881 newspaper article reported that two or three local men saw a creature that looked like a "huge monkey or baboon ... somewhat larger than a man." On October 3, 1894, while riding in the

YAHOO?

No one knows how the aborigines came to use the word "yahoo." In English it means a crude or stupid person. In the eighteenth and nineteenth century, English speakers sometimes used it to describe orangutans. Oddly enough, native inhabitants of the Bahamas also call their local apeman the "yahoo."

New South Wales bush in the middle of the afternoon, Johnnie McWilliams said he spotted a "wild man or gorilla" that stepped out from behind a tree, looked at him briefly, and dashed for a wooded hillside a mile away. Judging McWilliams a "truthful and manly fellow" when recording his story, the *Queanbeyan Observer* of November 30 also noted that "for many years there have been tales of trappers coming across enormous tracks of some unknown animals in the mountain wilds around Snowball."

Around the turn of the century, Joseph and William Webb were camped on a range in New South Wales. There they reportedly fired on a frightening-looking apelike creature that left "footprints, long, like a man's, but with longer, spreading toes"; its stride was also greater than a man's. According to John Gale in *An Alpine Excursion* (1903), the two found "no blood or other evidence of their shot having taken effect." And on August 7, 1903, the *Queanbeyan Observer* printed a letter from a man who claimed to have witnessed the killing of a yahoo by aborigines.

In the *Sydney Herald* of October 23, 1912, cattleman and poet Sydney Wheeler Jephcott wrote about the remarkable adventure of his neighbor, George Summerell. While riding his horse between Bombala and Bemboka around noon on October 12, Summerell came upon "a strange animal, which, on all fours, was drinking from the creek. As it was covered with gray hair, the first thought that rose to Summerell's mind was: 'What an immense kangaroo.' But hearing the horse's feet on the track, it rose to its full height, of about 7 ft., and looked quietly at the horseman. Then stooping down again, it finished its drink, and then, picking up a stick that lay by it, walked steadily away up a slope to the ... side of the road, and disappeared among the rocks and timber 150 yards away."

Jephcott added, "Summerell described the face as being like that of an ape or man, minus forehead and chin, with a great trunk all of one size from shoulder to hips, and with arms that nearly reached to its ankles."

After hearing his neighbor's report, Jephcott himself rode to the site. There he found several footprints that supported the story and found that "the handprints where the animal had stooped at the edge of the water" were "especially plain." Jephcott observed that the hands were humanlike, except that the little fingers were bent away "much like the thumbs." He found the feet "enormously long and ugly"—and with only four toes! These toes were long and flexible. "Even in the prints which had sunk deepest into the mud there was no trace of the 'thumb' of the characteristic ape's 'foot'," he noted.

Besides the fresh foot- and hand-prints, Jephcott could make out older ones; the animal had crossed that way before. Wanting to make casts of the tracks, the cattleman returned two days later with plaster. He gave his casts of two footprints and one handprint to a Professor David at the local university. While admitting that the copies were less than perfect, David thought that "any reasonable being will be satisfied by the inspection of these three casts that something quite unknown and unsuspected by science remains to be brought to light."

From Yahoo to Yowie

In more recent times in Australia, the word "yowie" has been used to describe huge, shaggy, man-like creatures. Like yahoo sightings, reports of yowies seem to occur almost always in the south and central coastal regions of New South Wales and Queensland's Gold Coast. The terms *yahoo* and *yowie* may well describe the same mysterious creature.

In any case, reports of yowies or yahoos span the entire twentieth century. In 1971 a team of Royal Australian Air Force surveyors landed in a helicopter on top of unclimbable Sentinel Mountain and were astonished to find huge manlike tracks (though too large for a man) in the mud. On April 13, 1976, in Grose Valley near Katoomba, New South Wales, five backpackers reported coming upon a bad-smelling, five-foot-tall yowie—a female, judging from its large breasts. And on March 5, 1978, a man cutting trees near Springbrook on the Gold Coast reported hearing what sounded like a grunting pig. He went into the forest looking for it but instead spotted "about 12 ft. in front of me, ... this big black hairy man-thing. It looked more like a gorilla than anything. It had huge hands, and one of them was wrapped around a sapling .... It had a flat black shiny face, with two big yellow eyes and a hole for a mouth. It just stared at me, and I stared back. I was so numb I couldn't even raise the axe I had in my hand."

"It had a flat black shiny face, with two big yellow eyes and a hole for a mouth. It just stared at me, and I stared back."

Rex Gilroy, who formed the Yowie Research Center in the late 1970s, claims to have collected over 3,000 reports.

Of course, none of this has changed the minds of Australian scientists. As one clearly put it, "The first and only primates to have lived in Australia were human beings." Graham Joyner, however, who has written much on the mystery, suggests that the "yahoo was an undiscovered marsupial [pouched mammal] of roughly bear-like conformation, which was referred to ... throughout most of the 19th and 20th centuries.... The Yowie, on the other hand, is a recent fiction."

Sources:

Bord, Janet, and Colin Bord, *Alien Animals,* Harrisburg, Pennsylvania: Stackpole Books, 1981.

Cryptozoology magazine, the following issues: 3, 1984, pp. 55-57; 4, 1985, pp. 106-112; 5, 1986, pp. 47-54; 6, 1987, pp. 124-129; 8, 1989, pp. 27-36; 9, 1990, pp. 41-51 and 116-119.

Extinction Reconsidered

- LIVING DINOSAURS

- MOKELE-MBEMBE

- PTEROSAUR SIGHTINGS

- THYLACINE

- THE MYSTERY OF THE SIRRUSH

- PALUXY FOOTPRINTS

Extinction Reconsidered

LIVING DINOSAURS

Do dinosaurs still exist? Does the question sound ridiculous? Indeed, most scientists believe—and we have all been taught—that these giant reptiles became extinct some 65 million years ago. Still, dinosaur sightings in remote regions of the world *are* reported from time to time! A handful of scientists, explorers, and nature writers have tried to make sense of these "unbelievable" accounts and, where possible, investigate them.

Much of the investigation has centered on a legendary creature generally referred to as **mokele-mbembe** and described as a sauropod-like animal. (Sauropods were huge plant-eating dinosaurs with long necks and tails, small heads, bulky bodies, and stumplike legs; *Diplodocus, Apatosaurus [Brontosaurus],* and *Brachiosaurus* were sauropods.) The first printed mention of the huge, plate-shaped footprints linked with the beasts appeared in a 1776 history of French missionaries in west-central Africa. In the next two centuries missionaries, colonial officials, hunters, explorers, and natives would give remarkably similar descriptions of the animals that supposedly made those tracks.

All sighting reports in recent years have come from the swampy, remote Likouala region of the Congo, an area in central Africa on both sides of the Congo river. In 1980 and 1981 University of Chicago biologist Roy P. Mackal led two expeditions there, the first in the company of herpetologist (reptile and amphibian scientist) James H. Powell, Jr., who had heard mokele-mbembe stories while doing crocodile research in west-central Africa. Neither expedition had a sighting, though Mackal and his companions did interview a number of native witness-

Engraving of an ichthyosaur and a plesiosaur, from Louis Figuier, *The World before the Deluge,* 1865.

es. The creatures, greatly feared, were said to live in the swamps and rivers. A band of Pygmies had supposedly killed one at Lake Tèle around 1959.

Though Mackal's expeditions were not able to reach remote Lake Tèle, a competing group headed by American engineer Herman Regusters successfully made the trip. Regusters and his wife, Kia Van Dusen, claimed that they saw huge, long-necked animals several times, both in the water and in the swampy areas around the lake. Congolese government biologist Marcellin Agnagna, who was a member of Mackal's second expedition, also arrived in the area in the spring of 1982 and reported a single sighting. Both the Regusters and Agnagna said that camera problems kept them from photographing the fantastic animals. Three more expeditions to the area, one English and two Japanese, brought no new sightings.

Other Dinosaurs in Africa

While Regusters was at Lake Tèle he heard a strange story. The local people told him that a few months earlier, in February 1981, the bodies of three adult male elephants had been found floating in the water. The cause of death seemed to be two large puncture wounds in the stomach area of each. These were not bullet holes, and the elephants still had their tusks, suggesting that poachers had not killed them. The natives blamed the deaths on a mysterious horned creature that lived in the nearby forests.

They called the mysterious creature *emela ntouka,* "killer of elephants." Nearly every report described it as the size of an elephant—or larger—with heavy legs that supported the body from beneath (not from the side, as in a crocodile) and a long, thick tail. Its face was like that of a rhinoceros, with a single horn attached to the front of its head. Comfortable in water or on land, it was a plant-eater, but it nonetheless killed elephants and buffalos with its great horn. In his 1987 book *A Living Dinosaur?,* Mackal suggested that such an animal, if it existed, was likely to be a kind of prehistoric rhinoceros or a horned dinosaur, like the triceratops.

Mackal also collected a handful of hazy reports about *mbielu mbielu mbielu,* "the animal with planks growing out of its back"—said to look like a stegosaur. Sightings of *nguma monene,* a huge serpent-like reptile with a sawlike ridge along its back and four legs along its sides, proved to be more credible. Among the witnesses of this animal was American missionary Joseph Ellis, who, in November 1971, said he saw such a creature leave the Mataba River and disappear into the tall grass. While Ellis did not get a complete look at the animal—he did not see its head and neck—he guessed from the portions of the body he observed above the water that it was over 30 feet long!

Ellis knew the animals of the Congo well and was positive that the creature could not have been a crocodile. Reports from native witnesses, which did include descriptions of the head and a long tail, suggested to Mackal that "we are dealing with a living link between lizards and snakes," perhaps a "lizard type ... from a primitive, semi-aquatic group known as dolichosaurs."

In 1932 biologist Ivan T. Sanderson and animal collector W. M. (Gerald) Russell had a strange and frightening experience in the Mamfe Pool, part of the Mainyu River in western Cameroon. The two men, with native guides, were in separate boats passing clifflike river banks dotted with deep caves when they suddenly heard ear-shattering roars—as if huge animals were fighting in one of them.

Swirling currents sucked both boats near the thundering cave's opening. At that point, Sanderson would recall, there "came another gargantuan gurgling roar and something enormous rose out of the

Nearly every report described it as the size of an elephant—or larger—with heavy legs that supported the body from beneath (not from the side, as in a crocodile) and a long, thick tail.

water, turned it to sherry-colored foam and then, again roaring, plunged below. This 'thing' was shiny black and was the *head* of something, shaped like a seal but flattened.... It was about the size of a full-grown hippopotamus—the head, I mean."

REEL LIFE

Baby ... Secret of the Lost Legend, 1985.

A sportswriter and his paleontologist wife risk their lives to reunite a hatching brontosaurus with its mother in the African jungle.

Jurassic Park, 1993.

Author Michael Crichton's spine-tingling thriller about dinosaurs genetically cloned from prehistoric DNA. All is well until the creatures escape from their pens. Violent, suspenseful, and realistic, with gory attack scenes, this film became the highest-grossing movie of all time.

The Land before Time, 1988.

Lushly animated adventure film about five orphaned baby dinosaurs who band together to try to find the Great Valley, a paradise where they might live safely.

The Lost World, 1992.

A scientific team ventures into uncharted jungles where they find themselves confronted by dinosaurs and other dangers. Based on the story by Sir Arthur Conan Doyle, and a remake of the groundbreaking 1925 silent film of the same name.

Sanderson and Russell decided not to stick around to observe anything more. Upstream they found big tracks that could not have been left by a hippopotamus—because hippos did not live in the area; natives said that the awful creatures had killed them all. Still, the strange animals were not flesh-eaters but fed instead on the liana fruits that grew along the rivers. According to Sanderson, the natives called the creatures *m'kuoo m'bemboo.*

If the part of the animal the explorers saw really was its head, then the creature was probably not the sauropodlike mokele-mbembe (sauropods have small heads). Mackal found during his own expeditions 50 years later that some native peoples used the same words to describe any large, dangerous animal living in rivers, lakes, or swamps.

Dinosaurs in South America

In his 1912 novel *The Lost World,* Sir Arthur Conan Doyle wrote of a band of hardy English explorers who discovered an isolated area in South America's Amazon River basin where prehistoric monsters lived on millions of years past their time. Considering that the tale has been a favorite of readers the world over for decades, it is surprising that so few claims of dinosaur sightings have come from South America in real life.

But one account was published in the *New York Herald* of January 11, 1911. It was written by a German named Franz Herrmann Schmidt, who claimed that one day in October 1907, he and his companion Capt. Rudolph Pfleng and their Indian guides entered a valley of swamps and lakes in a remote region of Peru. There they discovered some strange, huge tracks made by unknown animals, along with crushed trees and plants. They also found the area odd because it lacked the usual alligators, iguanas, and water snakes.

Despite the guides' fear, the expedition camped in the valley that night. The next morning the explorers got back into their boat and continued their search for the mystery animals. Just before noon they found fresh tracks along the shore. Pfleng declared that he was going to follow them inland, regardless of the danger. Then they heard the screams of a group of monkeys that were gathering berries in some nearby trees.

According to Schmidt: "[A] large dark something half hidden among the branches shot up among [the monkeys] and there was a great commotion." The frightened guides quickly paddled the boat away from shore, and while Pfleng and Schmidt could see very little then, they were able to hear "a great moving of plants and a sound like heavy slaps of a great paddle, mingled with the cries of some of the monkeys moving rapidly away from the lake." Then there was silence.

After about ten minutes, the plants near the lake began to stir again and the expedition members finally saw "the frightful monster." Unbothered by their presence, it began to enter the water and came to within 150 feet of them. The creature was gigantic; Schmidt figured it was "35 feet long, with at least 12 of this devoted to head and neck." Its head was "about the size of a beer keg and was shaped like that of a tapir, as if the snout was used for pulling things or taking hold of them." Its neck was thick and snakelike, and instead of forelegs it had "great heavy clawed flippers." Schmidt observed that its "heavy blunt tail" was covered with "rough horny lumps"; in fact, the entire surface of the creature's body was "knotted like an alligator's side."

Pfleng and Schmidt began shooting at the creature with their rifles. The bullets seemed to annoy it, but they drew no blood—the one that hit the creature's head flew off as if it had met with solid rock. Seven shots were fired, each reaching its target. In order to escape them, the creature suddenly plunged underwater and nearly overturned the explorers' boat! The monster reappeared some distance away. Schmidt reported that "after a few seconds' gaze it began to swim toward us, and as our bullets seemed to have no effect we took flight in earnest. Losing sight of it behind an island, we did not pick it up again and were just as well pleased."

The creature was gigantic; Schmidt figured it "was 35 feet long, with at least 12 of this devoted to head and neck."

Giant Footprints

According to Schmidt's records, the rest of the expedition's experiences along the Solimes River were ordinary. But because Pfleng died a few months later of fever, the incredible account of the mystery beast could not be supported by a second witness. Still, Schmidt's report was not the only one describing a huge swamp-dwelling creature in remote South America. In the early twentieth century, Lieutenant Colonel Percy H. Fawcett, who surveyed the jungle for Britain's Royal Geographical Society, wrote that natives had told him about "the tracks of some gigantic animal" seen in the swamps along the part of the Acre River where Peru, Bolivia, and Brazil meet—a few hundred miles from the site of Schmidt and Pfleng's strange experience. But the natives admitted that they had never actually seen the creature that left the prints. Fawcett also noted that farther south, along the Peru-Bolivian border, huge tracks of an unknown animal had been found.

Even Stranger Dinosaur Sightings

While the notion of dinosaurs still living in deepest Africa and South America is, at least, somewhat plausible, the presence of such creatures in the United States or Europe is all but impossible. Still, such sightings have been reported!

In a letter published in the August 22, 1982, issue of *Empire Magazine,* Myrtle Snow of Pagosa Springs, Colorado, wrote that in May 1935, when she was three years old, she saw "five baby dinosaurs" near her hometown. A few months later a local farmer shot one after it took some of his sheep. "My grandfather took us to see it the next morning," she said. "It was about seven feet tall, was gray, had a head like a snake, short front legs with claws that resembled chicken feet, large stout back legs and a long tail."

And these were not her only sightings! There were two more. In 1937 Snow saw another dinosaur in a cave, but this time it appeared to be dark green. And on the evening of October 23, 1978, in a driving rain, she spotted one crossing a field not far from where she had had her 1937 sighting.

In another case, in 1934, a South Dakota farmer claimed that a giant, four-legged reptile had forced his tractor off the road before disappearing into nearby Campbell Lake. Investigators found huge tracks on the shore. In addition, even before the sighting was reported, sheep and other small animals had been mysteriously vanishing.

Finally, a man reported being attacked by a "15-foot reptile, like a dinosaur," at Forli, Italy, in December 1970. Fifty miles northwest of

A South Dakota farmer claimed that a giant, four-legged reptile had forced his tractor off the road before disappearing into nearby Campbell Lake.

Dinosaur attack scene in 1933 horror film classic *King Kong*.

there, in June 1975, a monster appeared in a tomato field near Goro and badly frightened a farmer named Maurizio Tombini, who news accounts said had "a reputation for seriousness." The monster, according to news reports, measured about 10 feet long, had legs, and its feet left remarkable tracks. Tombini compared it to a "gigantic lizard" and denied that it was a crocodile. According to police, several other people also reported sightings. These witnesses noted that the creature had a wolflike howl.

Sources:

Heuvelmans, Bernard, *On the Track of Unknown Animals,* New York: Hill and Wang, 1958.
Ley, Willy, *Exotic Zoology,* New York: Viking Press, 1959.
Mackal, Roy P., *A Living Dinosaur?: In Search of Mokele-Mbembe,* New York: E. J. Brill, 1987.
Mackal, Roy P., *Searching for Hidden Animals,* Garden City, New York: Doubleday and Company, 1980.
Sanderson, Ivan T., *More "Things",* New York: Pyramid Books, 1969.

MOKELE-MBEMBE

The fascinating story of mokele-mbembe—pronounced "mo-kay-lee mmmbem-bee," a Lingala word meaning "one who stops rivers"— began in print, at least, in 1776. A book written by Abbe Lievain Bonaventure Proyart about French priests trying to bring Catholicism to the people of west-central Africa also contained a great deal of careful information about the animals of the region. It told of a startling discovery in the forest, where priests noted tracks left by an animal "which was not seen but which must have been monstrous": each clawed print measured about three feet across.

SAUROPODS

Sauropods were huge plant-eating reptiles with long necks and tails, small heads, bulky bodies, and stumplike legs; *Diplodocus, Apatosaurus (Brontosaurus)*, and *Brachiosaurus* were sauropods.

In the twentieth century Bernard Heuvelmans, the founder of cryptozoology (the study of unknown or unexpected animals), would figure that an animal that made such tracks must have been about the size of a hippopotamus or elephant. What could such an animal have been? (Or still be?) Legend—backed up by a number of eyewitness sightings—points to a sauropod, a type of dinosaur thought to have become extinct some 65 million years ago!

In the early 1870s Alfred Aloysius Smith, a young Englishman who worked for a British trading firm, was sent to the French African settlement of Gabon. In the course of his work, he traveled up and down the Ogooue River. Years later South African novelist Ethelreda Lewis would help him record his adventures in what would become the much-read book *Trader Horn,* published in 1927. In it he referred to mysterious animals that lived in the African swamps and rivers. "I've seen the *Amali's* footprints," he recalled. "About the size of a good frying pan ... and three claws instead o' five." In 1909, while on a westward journey along the rivers of central Africa, Lieutenant Paul Gratz, too, wrote of a scaleless, clawed swamp creature that frightened natives living near Lake Bangweulu in northern Rhodesia (now Zambia).

"Half Elephant, Half Dragon"

Carl Hagenbeck, a world-famous animal collector, was the first to bring wide attention to the subject. In his 1909 autobiography, *Beasts and Men,* he recalled how he had come to learn of the creature. At separate times an employee on a collecting expedition, Hans Schomburgh, and an English big-game hunter had told him about a "huge,

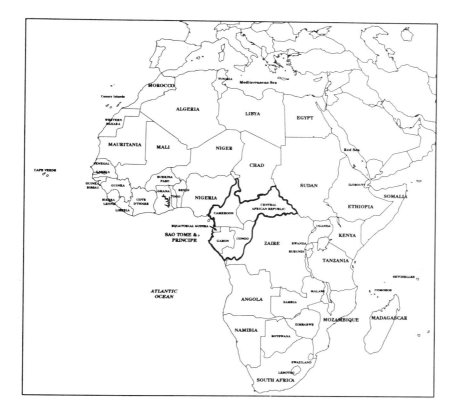

A map of Africa: Many mokele-mbembe sightings took place in central African countries such as the People's Republic of the Congo, Cameroon, Central African Republic, and Gabon.

monster, half elephant, half dragon," said to live in central Africa. And before that, collector and naturalist Joseph Menges had informed him of similar reports from natives describing what sounded to Hagenbeck like "some kind of dinosaur, seemingly akin to the brontosaurs." So fascinated was Hagenbeck, in fact, that he sent an expedition to the huge, swampy region where the beast reportedly lived. Disease and hostile natives kept the investigators from reaching the location, though, and Hagenbeck had to abandon his search.

Schomburgh told Hagenbeck that at Lake Bangweulu in Northern Rhodesia (now Zambia) he had been surprised to find no hippopotamuses. His native guides explained that this was because a strange animal that lived in the lake had killed the hippos. Five hundred miles to the west, in the Dililo marshes, he heard descriptions of a similar creature.

These accounts caused quite a stir in Africa, Europe, and the Unites States. Newspapers repeated the stories, with opinions ranging from harsh doubt to breathless excitement. The sensation faded after a few months, though, and in 1911, in a geographical journal, Northern

Rhodesian colonial official Frank H. Melland made this statement: "I have never heard so much as a rumor of any animal that could be supposed to resemble a brontosaurus, or a dinosaur, which has been reputed to inhabit these swamps."

Then, in 1913, the German government sent Captain Freiherr von Stein zu Lausnitz to survey colonial Cameroon, which bordered the African Congo. In his official report, not published until some years later, he noted that in areas of the lower Ubangi, Sanga, and Ikelemba rivers in the Congo, many people—including experienced hunting guides who knew the local wildlife well—spoke of something called mokele-mbembe. Von Stein wrote: "The animal is said to be of a brownish-gray color with a smooth skin, its size approximately that of an elephant, at least that of a hippopotamus. It is said to have a long and very flexible neck.... A few spoke about a long muscular tail like that of an alligator. Canoes coming near it are said to be doomed; the animal is said to attack vessels at once and to kill the crews but without eating the bodies."

> "The animal is said to be of a brownish-gray color with a smooth skin, its size approximately that of an elephant, at least that of a hippopotamus."

Von Stein added that the creature reportedly lived in caves at sharp bends along the river. "It is said to climb the shore even at daytime in search of food; its diet is said to be entirely vegetable.... The preferred plant was shown to me; it is a kind of liana with large white blossoms, with a milky sap and applelike fruits. At the Ssombo river I was shown a path said to have been made by this animal in order to get at its food. The path was fresh and there were plants of the described type nearby." Because there were so many tracks of other animals on the path, however, Stein could not make out those of the creature.

During the next few decades, two brontosaur-hunting expeditions were launched, but without much success. In a 1938 exploration, German Leo von Boxberger collected many mokele-mbembe reports but lost them in an attack on him and his group by unfriendly natives.

Outside the regions where the creatures reportedly lived, mokele-mbembe would have been forgotten had it not been for the continued interest of cryptozoologists Willy Ley, Ivan T. Sanderson—who wrote about them in a 1948 *Saturday Evening Post* article—and Heuvelmans. Heuvelmans included an entire chapter on the beasts in his 1958 book *On the Track of Unknown Animals.*

Into the Heart of Darkness

In the 1960s a young herpetologist (reptile and amphibian scientist) named James H. Powell, Jr., became interested in Africa's mystery

Extinction Reconsidered

A representation of what a mokele-mbembe might look like.

sauropod after reading about it in the writings of Ley, Sanderson, and Heuvelmans. While conducting field research on rain-forest crocodiles in west-central Africa in 1972, he tried to enter the People's Republic of the Congo to investigate the mokele-mbembe puzzle for himself. He was not allowed to enter the country; after four years of trying, Powell gave up and set off for Gabon and the remote regions where *Trader Horn* author Smith had seen tracks of strange animals so many decades before.

Natives Provide Strong Evidence

There Powell eventually found a witness who had seen the mystery beast. Without being asked, the witness mentioned the plant that, according to most reports, comprised the creature's diet. Shown pictures of different animals, the witness pointed to one of a diplodocus. Picture tests given to other native observers brought the same identification. In early 1979 Powell returned to the area to gather additional testimony from his first witness and from others.

A third expedition was launched in February 1980. This time Roy P. Mackal, a University of Chicago biologist with a keen interest in

Map of People's Republic of the Congo and surrounding areas.

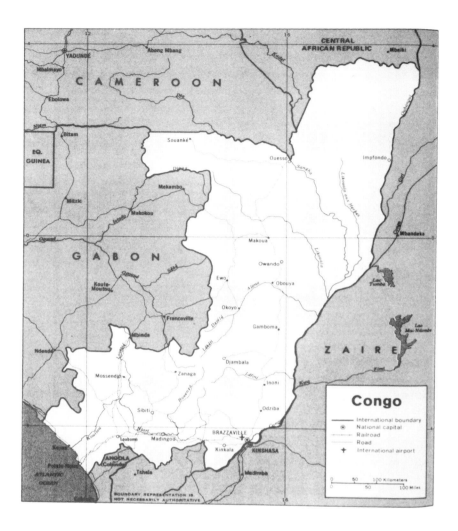

strange animal reports, went with Powell. The two focused their explorations on the northern Congo, between the Sanga and Ubangi rivers. Mackal thought that growing human traffic on the waterways might have pushed mokele-mbembe into the remote Likoula region—mostly swamp and rain forest and still largely unexplored.

The two scientists journeyed into the great swamp but observed nothing. They decided to use what little remained collecting the testimony of witnesses. American missionary Eugene Thomas, who had lived on the Ubangi River for many years and had often heard mokele-mbembe stories, was able to direct them to several native observers. The descriptions given were remarkably similar: animals 15 to 30 feet

Roy P. Mackal

(1925-)

Roy P. Mackal is a well-respected figure in cryptozoology. After serving in World War II, Mackal took his Ph.D. from the University of Chicago, where he spent his professional life until retirement in 1990.

Mackal taught and conducted research in the fields of biology and chemistry and made a number of major discoveries in viral research and genetic engineering. But Mackal has had many interests. An inventor and engineer with several patents to his credit, he is also the editor and publisher of a journal dedicated to the study of Liberian stamps and is writing a biography of Victorian actress Maude Adams.

He is best known, though, for his cryptozoological investigations and writings. Between 1965 and 1975 he was the scientific director of the Loch Ness Phenomena Investigation Bureau. His 1976 book *The Monsters of Loch Ness* is considered a cryptozoological classic.

In 1980 Mackal published a book intended for a popular audience, called *Searching for Hidden Animals*. After two expeditions to the Congo to pursue reports of mokele-mbembe, he wrote *A Living Dinosaur?* Since then he has participated in an expedition that investigated pterosaur sightings in southwest Africa. Mackal was also a cofounder of the International Society of Cryptozoology and has served as its vice president since 1982.

long, most with snakelike heads and necks and long, thin tails. Their bodies were bulbous, like that of a hippo. The largest specimens were "the size of a small elephant." Mackal and Powell were told that the beasts had "stubby legs, and that the hind feet [had] three claws." The creatures were described as hairless. Again, witnesses pointed to a sauropod when shown pictures of different animals for comparison.

The natives said that the animals, while they did not eat flesh, would overturn canoes that went into the rivers, lakes, and streams where they lived and that the local people would kill them. Mackal and Powell learned of a mokele-mbembe killing, sometime around 1959, after Pygmies trapped one in a channel going into Lake Tele. Later, it was reported, they ate its flesh—and all of them died.

In late 1981 Mackal returned to the area, this time with ecologist J. Richard Greenwell, Congolese government biologist Marcellin Agnagna, and others. They interviewed more native witnesses—though this was made difficult at times by the widespread belief that anyone who talked about mokele-mbembe would die soon afterward. Expedition members did not see any of the creatures themselves, although one time, as they were rounding a curve in the Likouala River, they heard a great "plop" sound. Then a large wake or track was seen on the surface of the water, suggesting that a large animal had gone—and remained—below. Hippopotamuses did not live in the area. Residents had told expedition members that mokele-mbembe were often seen at sharp bends in the rivers where the water was deepest.

The expedition members were also shown a trail of freshly broken branches through which an animal between five and seven feet high had passed on its way to a pool where a mokele-mbembe was said to live. There were 12-inch round tracks leading to the pool, but nowhere were there any tracks leading out of it!

Unable to break through the thick swamp growth, the expedition did not make it to the shores of Lake Tele. Two later expeditions claimed to do so, however, and in each case reported sightings of mokele-mbembe.

Sightings by Scientists

Around the time of the second Mackal expedition, American engineer Herman Regusters led a group—with much difficulty—to Tele. There members of the group camped for well over two weeks, during which they reported several sightings of the creature's head and neck poking out of the water. Once they heard a "low windy roar" on the lake, which "increased to a deep-throated trumpeting growl." Then Regusters and his wife, Kia Van Dusen, claimed that they saw a huge creature moving through the swamp along the lake's edge. Another time Regusters observed it or a similar animal through binoculars. It had a slender, eight-foot neck, a small head, and 15 feet of back;

It had a slender, eight-foot neck, a small head, and 15 feet of back; Regusters guessed that with its tail (which he did not see), the creature would measure "30 to 35 feet long."

Regusters guessed that with its tail (which he did not see), the creature would measure "30 to 35 feet long." However, intense humidity caused the expedition's cameras to fail, so there were no photographs of these remarkable sights.

In April 1983 Agnagna led an all-Congolese government expedition to Tele. On the morning of May 1, he and two local villagers set out to film animal and bird species in the forest surrounding the lake. One of Agnagna's companions fell into a muddy pool and went to the lake to wash himself. There he spotted something in the water and called for Agnagna, who waded far out until he, too, could see the large and remarkable animal about 1,000 feet away. The creature resembled a mokele-mbembe.

Agnagna carefully watched the animal for the next 20 minutes. He could make out a wide back, long, thin neck, and small head. This portion measured about 18 feet; with a tail (which the biologist guessed was under water), the animal would have been a good deal longer. The animal finally disappeared below the lake's surface. While he took photographs of the creature, Agnagna wrote that "the emotion and alarm at this sudden, unexpected event" must have caused him to set his camera incorrectly, a mistake he discovered only later, when the film was processed. He claimed that no photographs survived.

Yet at an International Society of Cryptozoology meeting in Paris in 1984, Agnagna gave a slightly different version of the sighting and his camera problems in a formal paper and interviews. And he produced snapshots of the lake creature taken with a still camera. These photographs showed only a small, distant, unidentifiable image—which is not surprising considering how far Agnagna was from it. The biologist never explained why he had changed his story.

Operation Congo

Other mokele-mbembe expeditions took place in the next few years in an exploration effort called Operation Congo. And in one of these, Agnagna's behavior was again called into question. Hired by four eager young Englishmen to lead them to Lake Tele in mid-1986, the biologist was accused of stealing the group's film and supplies (a matter which eventually ended up in court) and causing other disturbances. In addition, no sightings of mokele-mbembe were made.

Agnagna accompanied two Japanese expeditions in September 1987 and the spring of 1988. The second group made it to Tele, where they saw nothing out of the ordinary. Like other Operation Congo members before them, the Japanese interviewed a number of people who claimed to have seen the animals at some point in their lives.

A Not-So-Impossible Animal?

During his second expedition, Mackal collected more than 30 detailed reports of mokele-mbembe. He felt that they described "a small sauropod so well that I find it impossible not to accept the identification. Each of the reports was a firsthand, eyewitness account by informants from widely differing ethnic, cultural, religious, and geographical backgrounds."

With no body or bone or skin samples as evidence, those who have not seen mokele-mbembe for themselves must make their judgments from this sort of testimony. Although the existence of a modern-day sauropod is hard to believe, some think it is not altogether out of the question. While it is thought that dinosaurs became extinct—over a large span of time—some 65 million years ago, the Congo basin's geography and climate have remained the same for at least that long. Crocodiles, a close relative of dinosaurs, have survived all that time without much change. In other words, the survival of a small population of dinosaurs in a remote, stable, and suitable place is not totally crazy!

Sources:

Heuvelmans, Bernard, *On the Track of Unknown Animals,* New York: Hill and Wang, 1958.
Ley, Willy, *Exotic Zoology,* New York: Viking Press, 1959.
Mackal, Roy P., *A Living Dinosaur?: In Search of Mokele-Mbembe,* New York: E. J. Brill, 1987.
Mackal, Roy P., *Searching for Hidden Animals,* Garden City, New York: Doubleday and Company, 1980.
Sanderson, Ivan T., *More "Things",* New York: Pyramid Books, 1969.

PTEROSAUR SIGHTINGS

Pterosaurs, commonly and mistakenly referred to as pterodactyls, are extinct flying reptiles of the order Pterosauria, common in the Mesozoic era (from about 160 to 60 million years ago). They were not true flying reptiles, nor ancestors of birds; instead, they glided on winds and air currents—more like bats.

The wing of a pterosaur consisted of a thin layer of skin, stretched out along an *extremely* long fourth finger on the hand, then down to meet the lower leg. Early varieties of pterosaurs were generally small and had fully toothed jaws and long tails. Later forms were often large, had stumps for tails, and fewer teeth in jaws shaped like beaks. Since their legs were weak and their wings very heavy, pterosaurs were practically helpless on land. Eventually, the large reptiles were replaced by birds, which were better flyers.

Sightings

On January 11, 1976, two ranch hands near Poteet, Texas, just south of San Antonio, sighted a five-foot-tall birdlike creature standing in the water of a stock tank. "He started flying," witness Jessie Garcia reported, "but I never saw him flap his wings. He made no noise at all."

Around the same time, two sisters, Libby and Deany Ford, observed a "big black bird" near a pond northeast of Brownsville, near the Texas-Mexico border. "It was as big as me," Libby said, "and it had a face like a bat." Later, as the two girls looked through a book in an effort to identify the creature, they found out what it was.

> ### TYPES OF PTEROSAURS
>
> The pterodactylus was a small pterosaur with a 12-inch wingspan, a small number of teeth, and a stump of a tail. The last and greatest pterosaur, the pteranodon, had a wingspan of more than 20 feet; its beak had no teeth and there was a huge bony crest on the back of its head.

Driving to work on an isolated country lane southwest of San Antonio on the morning of February 24, three elementary school teachers saw a shadow cover the entire road. The object causing it, which was passing low overhead, looked like a huge bird with a 15- to 20-foot wingspan. "I could see the skeleton of this bird through the skin or feathers or whatever," witness Patricia Bryant said. According to observer David Rendon, "It just glided. It didn't fly. It was no higher than the telephone line. It had a huge breast. It had different legs, and it had huge wings, but the wings were very peculiar like. It had a bony structure, you know, like when you hold a bat by the wing tips, like it has bones at the top and in-between."

Having never seen anything like it, the three witnesses rushed to an encyclopedia as soon as they got to school. After some searching they found what they were looking for. They learned that the animal they had observed was not a mystery creature after all.

Pterosaurs in flight.

At 3:55 A.M. on September 14, 1982, ambulance technician James Thompson was driving along Highway 100, a few miles east of Los Fresnos, Texas, and midway between Harlingen and Brownsville. He suddenly spotted a "large birdlike object" pass low over the highway, 150 feet in front of him. Its tail was so strange-looking that it practically stopped him in his tracks. He hit the brakes, pulled his vehicle to the side of the road, and keenly stared at the strange object, which he at first had trouble believing was a living creature.

"I expected him to land like a model airplane," Thompson said. Then "he flapped his wings enough to get above the grass.... It had a black, or grayish, rough texture. It wasn't feathers. I'm quite sure it was a hide-type covering." Its thin body, which ended in a "fin," stretched over eight feet; its wingspan was five to six feet. The wings appeared to have "indentations." At the back of the head was a hump like a Brahma bull's. There was "almost no neck at all."

Later Thompson looked at books, trying to identify the "bird." Like the Ford sisters and the San Antonio teachers more than six years earlier, he had no real difficulty finding out what he had seen; the books indicated to him, as it had to the others, that he had seen an extinct pterosaur!

African "Breaker of Boats"

In the early twentieth century, a traveler and writer named Frank Melland worked for the British colonial service in Northern Rhodesia (now Zambia). While there he learned of a flying creature that lived along certain rivers. Called *kongamato*—"breaker of boats"—it was considered very dangerous. Natives said it was "like a lizard with membranous wings like a bat."

Melland wrote about the creature in his 1923 book *In Witchbound Africa*. From the natives he learned that its "wing-spread was from 4 to 7 feet across, [and] that the general color was red. It was believed to have no feathers but only skin on its body, and was believed to have teeth in its beak: these last two points no one could be sure of, as no one ever saw a kongamato close and lived to tell the tale." Amazingly, when Melland showed the local residents two books he had containing pictures of pterosaurs, "every native present immediately and unhesitatingly picked it out and identified it as a kongamato!"

The natives insisted that the flying reptile they described still existed. While unsure about this claim, Melland at least believed that the creature had lived sometime "within the memory of man." He concluded, "Whether it is scientifically possible that a reptile that existed in the mesozoic age could exist in the climatic conditions of to-day I have not the necessary knowledge to decide."

Looking back on his days as an African game warden, Colonel R. S. Pitman published his memoirs, *A Game Warden Takes Stock,* in 1942. In the book he recalled that when in Northern Rhodesia he had heard of a frightening mythical beast that brought death to those who looked at it. It was said to have lived (and possibly still lived) in the thick, swampy forest region near the Angola and Congo borders. What most fascinated Pitman about the mystery beast was that it was both batlike and birdlike—and gigantic—resembling the prehistoric creature known as the pterosaur.

Similarly, in his 1947 book *Witchcraft and Magic in Africa,* Frederick Kaigh referred to a spot on the "Rhodesian-Congo border near the north-eastern border of the Jiundu Swamp, a foetid [smelly], eerie place in which the pterodactyl is locally supposed to survive with spiritual powers of great evil."

According to Carl Pleijel of the Swedish Museum of Natural History, a sighting of such a pterosaur-like creature took place in Kenya in 1974. The witnesses were members of a British expedition, Pleijel told writer Jan-Ove Sundberg. Not long after this report, Sundberg also learned of a second sighting over a swamp in Namibia in late 1975 by an American expedition.

Namibia has been the source of other such reports. In the summer of 1988, cryptozoologist Roy P. Mackal traveled there with a small group of associates. He was particularly interested in an isolated private desert area where sightings of "flying snakes" continued to be reported. Mackal interviewed witnesses, who said that the animals indeed had wings—of 30 feet, no less—but no feathers. The creatures appeared to live in the caves and cracks in the many kopjes (grassland hills) that dotted the landscape. Expedition members found ostrich bones in almost unreachable spots atop kopjes, suggesting that kills had been carried there by flying creatures. One expedition member

What most fascinated Pitman about the mystery beast was that it was both batlike and birdlike— and gigantic— resembling the prehistoric creature known as the pterosaur.

A pterosaur in a scene from the movie *The Animal World.*

who stayed on after Mackal returned to the United States reported seeing one of the creatures from a thousand feet away. It was, he said, black with white markings and had huge wings, which it used to glide through the air.

Sources:

Clark, Jerome, and Loren Coleman, *Creatures of the Outer Edge,* New York: Warner Books, 1978.

Coleman, Loren, *Curious Encounters: Phantom Trains, Spooky Spots, and Other Mysterious Wonders,* Boston: Faber and Faber, 1985.

Heuvelmans, Bernard, *On the Track of Unknown Animals,* New York: Hill and Wang, 1958.

Mackal, Roy P., *Searching for Hidden Animals,* Garden City, New York: Doubleday and Company, 1980.

THYLACINE

One of cryptozoology's most interesting puzzles involves the thylacine, an Australian animal that has been officially declared extinct—yet may not be. The case of the thylacine is actually a mystery that has two parts.

A "Tasmanian Tiger"

The thylacine, a flesh-eating marsupial (pouched mammal), evolved on the Australian mainland late in the age of mammals. Though it looked like a mixture of a fox, a wolf, a tiger, and a hyena, it was actually related to the opossum (zoologists believe they share a common ancestor).

The male thylacine measured over six feet long between head and tail. Its head looked like that of a fox or a dog, but beginning mid-back and going all the way to the tail, it had tigerlike stripes (from which it got its popular nickname, "Tasmanian tiger"). Its long, bunched rear resembled a hyena's and ended in a stiff, unwagging tail. Its fur was coarse and a sandy-brown color. Females were a little smaller, but with twice the number of stripes, which started just behind the neck. They also had a pouch—as all female marsupials do—but it faced the rear, perhaps to protect its young as it moved through the undergrowth.

MARSUPIALS

A marsupial is any mammal of the order Marsupialia, which includes kangaroos, opossums, bandicoots, and wombats. The female of most species lacks a placenta—the organ that, in most mammals, joins the unborn young to the mother so the fetus can get nourishment. Instead of the placenta a marsupial mammal has an abdominal pouch called a marsupium. Located outside the mother's body, the marsupium contains milk glands and shelters the young until they are fully developed.

History

About 12,000 years ago, thylacines were driven off the mainland of Australia, probably when Indonesian sailors brought dingoes (a reddish-brown wild dog) over, which proved to be excellent hunters. At least that is what zoologists think, because no fossil record of thylacines on the mainland can be found from that time on. It is believed that the animal retreated to Tasmania. While now an island state off Australia's southeast coast, Tasmania then was connected to the continent by a land bridge. The first mention of a thylacine in print was in an 1805 Tasmanian newspaper, which called the animal "destructive."

Convinced that thylacines were behind the mass killings of sheep on ranches, officials began a campaign to wipe out the "tigers"

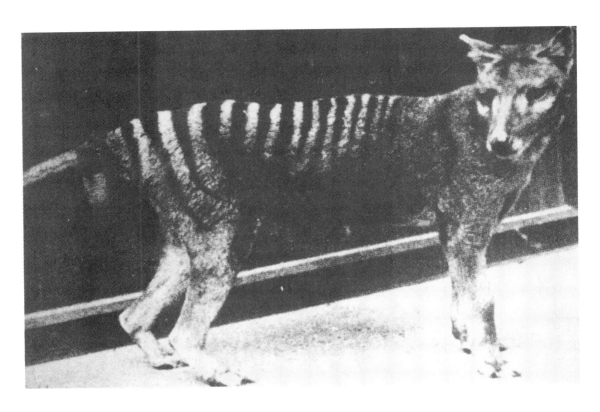

(although, according to modern experts, wild dogs and human rustlers were a far greater threat to livestock than the hated thylacines ever were). Both private companies and the government offered rewards for thylacine scalps—and the killings began. By the early twentieth century, the animals had become a rare sight. Bounty hunters were not their only enemies, though; a distemper virus destroyed many of them, and settlers were claiming more and more of their natural home. Still, armed human beings brought about the thylacine's ultimate extinction. "Farmers continued to see the creature as a menace," two historians of the thylacine wrote, "long after it was [incapable] of reproducing itself in any numbers."

According to the records, the last specimen for which a bounty was paid was killed in 1909. The last time a thylacine was shot was in 1930. And, captured in 1933, the last one on earth died on September 7, 1936, in Tasmania's Hobart Domain Zoo, just two months after the state passed a law declaring the thylacine a protected species. "Benjamin," as the one remaining thylacine was named, "was tame and could be patted," according to keeper Frank Darby, but "it was frequently morose and showed no affection."

The First Mystery

Only after the thylacine was gone for good did most Tasmanians begin to feel sorry for what they had done. In time, the Tasmanian coat of arms would proudly show two thylacines, and Australians would consider it their most beloved lost animal. Its disappearance would come to be viewed as a national tragedy.

Yet as little as a year after Benjamin's death and the thylacine's official extinction, Australia's Animals and Birds Protection Board sent two investigators into the mountains of northwestern Tasmania to see if, perhaps, a few thylacines remained. They returned with a handful of promising reports of sightings collected from residents of the area. While not hard proof, the accounts did encourage the board to set up further searches. And in 1938 an expedition found the first physical evidence: tracks with the thylacine's unusual five-toed front paws and four-toed hind paws.

World War II stopped further investigation, but in late 1945 a private expedition viewed a set of tracks and heard some sighting reports—though none of its members saw anything themselves. Then Australian wildlife experts did nothing about the thylacine question for a number of years. That is, until 1957, when zoologist Eric R. Guiler, chairman of the Animals and Birds Protection Board, went to Broadmarsh to check out a mysterious animal that was killing sheep. Guiler had no doubts that the tracks he saw were those of a thylacine. Convinced that the animal still lived, Guiler launched nine expeditions between 1957 and 1966, during which he gathered a great deal of evidence. Still, he could produce no body, nor did he have a personal sighting.

In 1968 other researchers created the Tiger Center, a place where witnesses could report sightings. Search parties continued to explore the bush (a wilderness region in Australia). And a late-1970s project sponsored by the World Wildlife Fund set up a number of automatic cameras at locations where witnesses said they had seen thylacines; bait was placed to lure the animals across an infra-red beam that would trigger a photograph. Nine different species of animals were caught on film, particularly the Tasmanian devil (an animal which had, from time to time, been mistaken for the thylacine). In his official report in 1980,

---

TASMANIAN DEVIL

The *Tasmanian devil* is a powerful, flesh-eating marsupial—or pouched mammal—found only on the island of Tasmania. (It was once found on mainland Australia as well.) It is the size of a large cat or badger, and its blackish fur is marked with white patches on the throat, sides, and rump. The animal has a huge appetite and often kills species larger than itself. It lives in burrows in rocky areas. Its evil-looking face and fierce snarl are believed to be the reasons behind its frightening name.

The World Wildlife
Fund set up a number
of automatic cameras
in an unsuccessful
attempt to catch
thylacines on film.

project leader Steven J. Smith of the National Parks and Wildlife Service (NPWS) concluded that thylacines were indeed extinct.

But by 1982, in a published survey of 104 sightings reported to the NPWS between 1970 and 1980, Smith had changed his mind. Along with coauthor D. E. Rounsevell, he felt that the best course of action in finding the few thylacines that might remain was to carefully study "the growing collection of reported sightings"—most clustered in the northern regions of the state, where most of the killings had taken place so many decades ago.

In the meantime, Guiler also conducted a hidden-camera operation, with the same poor results. Regardless, he remained convinced that thylacines still existed because of continued sightings and tracks.

Events in the thylacine puzzle took a new turn one rainy night in March 1982. An NPWS park ranger in a forested area of northwestern Tasmania awoke from a nap in the back seat of his car. He turned on his spotlight and shined it on an animal 20 feet away. He said it was a thylacine, "an adult male in excellent condition, with 12 black stripes on a sandy coat." However, the rain wiped out any tracks it might have made.

The NPWS did not make the sighting public until January 1984, hoping to keep away the curious, who might frighten the animal and whatever companions lived with it. Yet the NPWS announcement was not an official statement that the thylacine was no longer extinct. After all, agency personnel could not produce the animal. In addition, they were afraid of the problems an official statement might bring: what would happen to mining and timber companies the property of which was discovered to be the habitat of the thylacine? The endangered animals would have to be protected; would mineral and lumber rights—along with business profits and the tax money they provided the country—have to be relinquished?

Since then other expeditions have been launched, but Tasmania's thylacine has not been found. Doubters think that sighting reports are just wishful thinking or mistaken identifications of other animals, especially wild dogs. In any case, sightings continue. In 1991 as many as 13 accounts were reported; wildlife officials judged three of them "very good." Zoologist Bob Green remarked of thylacines: "They are extremely cunning animals. For every one that's seen, I believe they see a thousand humans. I have received samples of dung and footprints sent in by experienced bushmen who know what they have seen. I believe that the thylacine not only exists but is coming back strongly."

The Second Mystery

In 1981, following a number of sightings of an unusual animal in a southwestern area of Western Australia, the state hired native tracker Kevin Cameron to investigate. In due time Cameron would claim to have seen the animal himself, and he identified it as a thylacine.

Even those who seriously consider that thylacines may have survived in Tasmania have a hard time believing reports like this. Not only do fossils show that no thylacines lived on the mainland in the past 12,000 years, but there is also no proof that the animals were known to Australia's native inhabitants, the aborigines, or to the Europeans who settled the continent in the nineteenth century.

In 1951 a man from Dwellingup visited the Western Australian Museum in Perth, where he displayed photographs and casts of tracks of what he believed was a thylacine. But the staff zoologist there, Athol M. Douglas, rejected the man's eyewitness account and supporting evidence. Some years later other reports of a strange sheep-killing animal led Douglas into the bush, where he tracked and killed the culprit, a house dog gone wild. The experience only made him doubt thylacine reports further. Even so, over the years he was called in to examine the slaughtered bodies of kangaroos and sheep from time to time. And he was bothered that the animals appeared to have been slain in exactly the way that thylacines—and not dogs or dingoes—would kill their victims.

In February 1985 all of Douglas's doubts disappeared when tracker Cameron handed him five color photographs. They showed the side view of an animal burrowing at the base of a tree. Though its face was hidden in the brush, its striped back and long, stiff tail could only be that of a thylacine. The tracker could not tell Douglas where he had taken the pictures, but he did produce casts of the animal's prints. Douglas thought that Cameron's description of the creature's behavior matched similar accounts in scientific writings. Because Cameron could barely read, this information made the zoologist believe his witness even more.

Fake Photos

Still, Cameron was acting odd and secretive, and Douglas thought that something about the photographs might be fishy. With some difficulty, the zoologist got Cameron to give him permission to publish the pictures with an article he had written for the British magazine *New Scientist*. Only then did Cameron show him the original negatives of the film, which proved that the tracker had, indeed, been lying.

As Douglas later recalled in *Cryptozoology:* "The film had been cut, frames were missing.... There were no photographs of the animal bounding away. Furthermore, in one negative, there was a shadow of another person pointing what could be an over-under .12 shotgun. Cameron had told me he had been alone. It would have been practically impossible for an animal as alert as a thylacine to remain stationary for so long while human activity was going on in its vicinity." Also, the zoologist wondered why the animal's head was never shown.

New Scientist readers noticed inconsistencies with the photographs as well. For one thing, the animal did not move at all from one picture to the next. Furthermore, shadow patterns in the photographs showed that at least an hour, or even more, separated some shots from the rest. To critics this could only mean one thing—that, over time, Cameron had photographed a stuffed model of a thylacine.

But Douglas thought it more likely that one of the pictures, the first taken, showed a living thylacine about to be shot by the shadowy man pointing the gun. The zoologist suspected that the rest of the photographs were taken later, after the animal had been dead for some time and was stiff with rigor mortis. Douglas hoped that someone would find the animal's dead body, but nobody ever did. Perhaps that is because there is a $5,000 fine for anyone convicted of killing the (officially extinct) thylacine.

The Evidence Builds

In 1966 a Western Australia Museum team had found a thylacine's remains in a cave near Mundrabrilla Station. Carbon dating showed that the body was 4,500 years old (all other thylacine fossil remains found on mainland Australia had been there 12,000 years or longer!). And, according to zoologist Douglas, a possible mistake in dating (due to contamination from groundwater that soaked the thylacine body) could make the specimen much more recent than that! Compared to a dingo body that Douglas found in the cave in 1986 (which was hairless, dry, and odorless, and thought to be 20 years old at most), the thylacine remains were, according to the zoologist, "in a far superior state of preservation." To him, the condition of the thylacine's body suggested that it had died no more than a year earlier, probably less!

The state of Western Australia (which makes up about one-third of the Australian continent) is not the only part of the mainland to claim thylacine sightings. According to investigator Rex Gilroy, many reports of "large striped dog-like animals, possibly thylacines" have been recorded in other parts of the country. He added that "plaster casts have been made of tracks found on the mainland," and these "compare with others from Tasmania, leaving little doubt as to the animal's identity."

Thylacine sightings have also taken place in the mountain wilderness of the Namadgi-Kosciusco National Park, along the New South Wales-Victoria border. There ranger Peter Simon reported seeing such an animal in broad daylight for several seconds some 100 feet away. In

OFFICIALLY EXTINCT AND MISPLACED, TOO?

On April 7, 1974, at 3:30 A.M., Joan Gilbert spotted a "strange striped creature, half cat and half dog," as it passed in front of her car's headlights. "It was," she recalled, "the most peculiar animal I have ever seen. It had stripes, a long thin tail, and ... was as big as a medium-sized dog." When she looked through reference books at the library, she discovered that it was an animal she had never heard of before: a thylacine. Funny thing, though—the sighting did not take place in Tasmania. It did not take place in Western Australia, or Victoria, or New South Wales. It happened outside of Bournemouth, in England!

1990, when Graeme O'Neill wrote an article about the thylacine mystery for Melbourne's leading newspaper, *The Age,* he received many cards and letters from Victoria residents reporting their own quite believable—and remarkably similar—sightings of the animal.

To Australian writer Tony Healy, there is something downright spooky about mainland thylacines. He noted that the night before ranger Simon's sighting, his hunting dogs refused to get out of the truck after they and their master heard harsh, thylacine-like panting sounds in the bush. In 1982 a Western Australia farm couple who claimed to have lost livestock to thylacines told a Perth newspaper that a "prickly feeling" at the back of their necks was always their first warning that the animals were near.

Sources:

Douglas, Athol M., "The Thylacine: A Case for Current Existence on Mainland Australia," *Cryptozoology* 9, 1990, pp. 13-25.

Douglas, Athol M., "Tigers in Western Australia?," *New Scientist* 110,1505, April 24, 1986, pp. 44-47.

Guiler, Eric R., *Thylacine: The Tragedy of the Tasmanian Tiger,* Oxford, England: Oxford University Press, 1985.

Wilford, John Noble, "Automatic Cameras Stalk Tasmania's Rare Tiger," *New York Times,* May 27, 1980.

THE MYSTERY OF THE SIRRUSH

It has long been believed that the sirrush was an imaginary animal. Or was it?

Around 600 B.C., during King Nebuchadnezzar's reign in Babylonia, an ancient country in southwest Asia, an artist carved a series of images of three different animals in the huge archway of Babylon's grand Ishtar Gate and on the high walls surrounding the road that approached it. The three beasts depicted were the lion, the rimi (a now-extinct wild ox), and the sirrush, which looked like a dragon. It has long been believed that the sirrush was an imaginary animal. Or was it?

Fascinated by the "zoological puzzle" of the sirrush, writer Willy Ley described its appearance thus: "[It has] a slender body covered with scales, a long slender scaly tail, and a long slim scaly neck bearing a serpent's head. Although the mouth is closed, a long forked

A dragon. Mythological creature or surviving dinosaur?

tongue protrudes. There are flaps of skin attached to the back of the head, which is adorned (and armed) with a straight horn."

One of the books of the Bible related that Nebuchadnezzar's priests kept a "great dragon or serpent, which they of Babylon worshipped." And many years earlier, in the Old Testament's Book of Job, the sirrush may have been referred to by another name. The Bible described a "Behemoth" that ate grass and laid "under the shady trees, in the cover of the reed, and fens [low land covered with water or swamps]." The creature had mighty strength, its bones were "like bars of iron," and it had a "tail like a cedar."

The behemoth's identity has long puzzled Bible experts, who believe that Job was writing about a real animal. University of Chicago biologist Roy P. Mackal offered this idea: "The behemoth's tail is compared to a cedar, which suggests a sauropod. This identification is reinforced by other factors. Not only the behemoth's physical nature, but also its habits and food preferences are compatible with a sauropod's. Both live in swampy areas with trees, reeds and fens.

The modern discoverer of the Ishtar Gate, German archaeologist Robert Koldeway, seriously considered that the sirrush was a real animal. For he noted that unlike descriptions of other fantastic beasts in Babylonian art, images of the sirrush remained unchanged over centuries. Still, Koldeway felt that saurians did not live at the same time as human beings and that the Babylonians did not have the skills to reconstruct such animals from fossil remains.

Babylonian Dinosaur Sightings in the African Congo

But Babylonians were known to have reached the African Congo—the home of the **mokele-mbembe**—in their travels. Ley, Mackal, and noted zoologist Bernard Heuvelmans have all suggested that the Babylonians heard of such creatures while there, perhaps sighted them, or even took specimens home.

Some modern scholars, Adrienne Mayor for one, believe that ancient peoples did, in fact, know of, and had interest in, prehistoric animals. Mayor has attested, "Reliable ancient sources relate that, when fossils were discovered in antiquity, they were transported with great care, identified, preserved, and sometimes traded. Reconstructed models or the remains of 'unknown' species were displayed in Greece and Rome." If such was the case with the sirrush, however, the fossilized remains would have had to come from elsewhere, as dinosaur fossils have never been found in the Mesopotamian region.

Sources:

Heuvelmans, Bernard, *On the Track of Unknown Animals,* New York: Hill and Wang, 1958.
Ley, Willy, *Exotic Zoology,* New York: Viking Press, 1959.
Mackal Roy P., *A Living Dinosaur?: In Search of Mokele-Mbembe,* New York: E. J. Brill, 1987.
Mackal, Roy P., *Searching for Hidden Animals,* Garden City, New York: Doubleday and Company, 1980.

PALUXY FOOTPRINTS

In the late 1930s a field explorer for the American Museum of Natural History named Roland Bird made a discovery in the limestone bed of the Paluxy River near Glen Rose, Texas, that upset one of science's most basic concepts—how life began on earth. By this time most scientists believed in the theory of *evolution* formulated by Charles Darwin in the mid-nineteenth century, which states that complex types of life developed, over time, from lower, or more simple, forms. Scientists believed that dinosaurs and human beings did not exist on the earth at the same time, that man appeared 60 million years after dinosaurs became extinct. Yet Bird discovered fossil tracks of dinosaurs and—what looked like—humans together in the same Cretaceous rock, 100 million years old! It was true that the "human" prints were *very* large—15 to 20 inches in length and eight inches in width—but they did show believable insteps and heels.

Yet Bird discovered fossil tracks of dinosaurs and—what looked like—humans together in the same Cretaceous rock, 100 million years old!

Evolutionism versus Creationism

A second theory about how life began on earth is called *creationism*. Based word for word on the Bible's Book of Genesis, it states that all living things, as they now exist, were created by God all at once. The Paluxy tracks seemed to support the beliefs of the creationists because they showed that simple and complex life forms did exist together in the past; creationists also held that the tracks proved that the earth was not as old as most scientists believed, that dinosaurs died in the Great Flood (in 4000 B.C.) described in the Bible, and that—again, according to the Bible—giants had once walked the earth (also see entry: **The Search for Noah's Ark**). For more than four decades, evolutionists and creationists argued back and forth about the footprints, with evolution scientists feeling that the "human" tracks were carved in later or that they really belonged to dinosaurs, but time and wear had kept the reptiles' special toe marks from being preserved.

It was not until the 1980s that the puzzle was finally solved. Glen J. Kuban, a computer programmer from Ohio who had been studying the footprints on and off for four years, found faint colors in the Paluxy tracks limestone in the pattern of dinosaur toes. In other words, different rock materials from those making up the rest of the tracks had filled in the toe marks and later hardened. At first dinosaur experts didn't believe Kuban's findings because it was believed that bipedal (upright, two-legged) dinosaurs never pressed the full weight of the soles of their feet on the ground, walking, instead, on their toes. But after other paleontologists found the same color differences in similar tracks near Clayton, New Mexico, they accepted Kuban's explanation as fact. Kuban and a friend, Ronnie Hastings, invited leading creation scientists to the Paluxy site and persuaded them that these were dinosaur—and not human—footprints after all.

Sources:

Kitcher, Philip, *Abusing Science: The Case Against Creationism,* Cambridge, Massachusetts: The MIT Press, 1982.
Steiger, Brad, *Worlds Before Our Own,* New York: Berkley-Putnam, 1978.
Wilford, John Noble, "Fossils of 'Man Tracks' Shown to Be Dinosaurian," *New York Times,* June 17, 1986.

Other Fantastic Creatures

- REPTILE MEN

- MOTHMAN

- BLACK DOGS

- CRAWFORDSVILLE MONSTER

- ONZA

- RI

Other Fantastic Creatures

REPTILE MEN

The 1954 science-fiction film *The Creature from the Black Lagoon* featured a strange animal from the Amazon River that walked upright on two legs but had gills and scales. While reports of such "reptile men" are not common, they do pop up from time to time. As early as 1878 a creature measuring six feet five inches tall and covered with "fish scales" was described at a Louisville, Kentucky, theater sideshow as the "Wild Man of the Woods." While the creature was probably an actor in costume who was after the money of his foolish audience, residents in a nearby area did report seeing a bipedal (two-legged) "giant lizard" nearly a century later!

More puzzling was the November 1958 account of a Riverside, California, man. While driving near the Santa Ana River he was attacked by a creature with a head like a scarecrow, shiny eyes, and scales; it left long scratches on his windshield. As witness Charles Wetzel sped away, he hit the monster and drove over it. The following evening another driver in the area reported a similar experience.

Weirder still was the case of Mrs. Darwin Johnson of Evansville, Indiana. While swimming in the Ohio River on August 21, 1955, she was dragged under the water by a clawlike hand that gripped her knee. Every time she struggled to reach the surface the unknown assailant pulled her down again. The thumping sound she made while grasping a friend's inner tube finally scared the attacker away. While never actually seen, the creature left a green palm stain on Mrs. Johnson's knee and scratches severe enough to warrant a trip to the doctor.

The reptile man in the 1954 film *The Creature from the Black Lagoon.*

That same year, along the Miami River in Loveland, Ohio, a man driving home from work at 3:30 A.M. on May 25 came upon a strange scene. Parking his car, he watched three awful-looking creatures with lopsided chests, wide, lipless, froglike mouths, and wrinkles instead of hair on their heads; one held a device that emitted sparks, and an odd odor filled the air. The witness reported the incident to his local police chief.

Nearly 17 years later, at 1 A.M. on March 3, 1972, two Loveland police officers saw a similar creature: a four-foot-tall, frog-faced, two-legged beast with leathery skin. The monster jumped a roadside guard rail on its way down to the Little Miami River. About two weeks later one of the officers reported a similar sighting, with the creature lying by the side of the road before it crossed a guard rail. The policeman shot at it but missed. A local farmer also reported seeing the monster.

On August 19, 1972, at Thetis Lake in British Columbia, Canada, a silver creature emerged from the water to chase two young men from the beach. One of the witnesses received cuts on his hand from the six sharp points atop the monster's head. A few days later, another person got a look at the creature, also noting a sharp point on its head. In addition, the witness claimed it had a "scaly" human-shaped body, "monster face," and "great big ears."

The true identity of these reptile men have not been found to this day.

Sources:

Coleman, Loren, *Curious Encounters: Phantom Trains, Spooky Spots, and Other Mysterious Wonders,* Boston: Faber and Faber, 1985.

Keel, John A., *Strange Creatures from Time and Space,* Greenwich, Connecticut: Fawcett Books, 1970.

In March 1972 several badly frightened Loveland residents, including two police officers, reported encounters with a bizarre frog-faced biped.

MOTHMAN

Late on the evening of November 15, 1966, as they drove past an abandoned TNT factory near Point Pleasant, West Virginia, two young married couples spotted two large eyes, two inches wide and six inches apart, attached to something that was "shaped like a man, but bigger. Maybe six or seven feet tall. And it had big wings folded against its back." The eyes were "hypnotic," the witnesses agreed. When the creature started to move, the four panicked and sped away. But they saw the same or a similar monster on a hillside near the road! It spread its batlike wings, rose into the air, and followed the car—which by now was going 100 mph.

"That bird kept right up with us," Roger Scarberry, one of the group, said to investigator John A. Keel. "It wasn't even flapping its wings." The witnesses told local deputy sheriff Millard Halstead that it

A drawing of Mothman, based on eyewitness descriptions.

made a sound like a "record played at high speed or the squeak of a mouse." It followed them on Highway 62 right to the Point Pleasant city limits.

The two couples were not the only people to see the creature that night. Another group of four claimed to have seen it not once but three times! A third sighting took place that evening. At 10:30 P.M. Newell Partridge, a builder who lived outside Salem, West Virginia (about 90 miles from Point Pleasant), was watching television when suddenly the screen went blank. Then a "fine herringbone pattern appeared on the tube, and ... the set started a loud whining noise, winding up to a high pitch, peaking and breaking off.... It sounded like a generator winding up." Partridge's dog Bandit began to howl on the porch and continued even after the set was turned off.

Partridge stepped outside, where he saw Bandit facing toward the hay barn 150 yards away. "I shined the light in that direction," he told West Virginia writer Gray Barker, "and it picked up two red circles, or eyes, which looked like bicycle reflectors"—only much larger. Something about the sight deeply frightened him, for he was certain that they were not animal eyes.

The snarling Bandit, an experienced hunting dog, shot off toward the creature. Partridge called to him to stop, but the dog paid no attention. At this point the man went inside to get a gun but then decided not to go outside again. He slept that night with the weapon by his side. By the morning he realized that Bandit had not returned. And the dog had not shown up two days later when Partridge read a newspaper report of the Point Pleasant sightings.

One detail in the newspaper account particularly grabbed his attention: Roger Scarberry had stated that as they entered Point Pleasant's city limits, the two couples had seen the body of a big dog by the side of the road. And a few minutes later, on their way back out of town, the dog was gone. Partridge immediately thought of Bandit, who would never be seen again. All that remained of him were his prints in the mud. "Those tracks were going in a circle, as if that dog had been chasing his tail—though he never did that," his master recalled. "There were no other tracks of any kind."

Other Fantastic Creatures

And there seemed to be another connection between the two sightings. Deputy Halstead had experienced strange interference on his police radio when he approached the TNT factory. It was loud and sounded something like a record or tape played at high speed. He finally had to turn the radio off.

The next day, after a press meeting called by Sheriff George Johnson, the story was reported across the country. One newspaperman dubbed the creature "Mothman" after a villain on the *Batman* television series.

More Sightings

From that time to November 1967, a number of other sightings occurred. On the evening of November 16, 1966, for instance, three adults—one carrying an infant—were walking back to their car after visiting friends. Suddenly, something rose up slowly from the ground. One witness, Marcella Bennett, was so frightened that she dropped her baby. It was a "big gray thing, bigger than a man," and it had no head. But it did have two large glowing red circles at the top of its torso. As huge wings unfolded from behind it, Raymond Wamsley snatched up the child and guided the two women inside the house they had just left. It seemed that the creature followed them to the porch, because they could hear sounds there and, worse still, see its eyes peering through the window. By the time the police arrived, however, it was gone. Bennett was upset for weeks afterward and, like other Mothman witnesses, eventually sought medical attention.

John Keel, the main investigator of the Mothman sightings, wrote that at least 100 people had seen the creature. From their accounts he put together a description. According to reports, it stood between five and seven feet tall, was broader than a man, and walked in a clumsy, shuffling manner on humanlike legs. It made a squeaky sound. The eyes, which Keel said "seemed to have been more terrifying than the tremendous size of the creature," were set near the top of the shoulders. Its wings were batlike but did not flap when it flew. When it took off from the ground, it went "straight up, like a helicopter," according to one observer. Witnesses described its skin color as gray or brown. Two observers said that they heard a mechanical humming as it flew above them.

After 1967 Mothman sightings died away. (Only one later account, in October 1974 from Elma, New York, was reported.) Nonetheless, Keel did locate a woman who said that she had met such a creature on

And it seemed that the creature followed them to the porch, because they could hear sounds there and, worse still, see its eyes peering through the window.

a highway one evening in 1961, on the West Virginia side of the Ohio River. She told Keel: "It was much larger than a man. A big gray figure. It stood in the middle of the road. Then a pair of wings unfolded from its back, and they practically filled the whole road. It almost looked like a small airplane. Then it took off straight up ... disappearing out of sight in seconds."

REEL LIFE

Mothra, 1962.

Classic Japanese monster movie about an enraged giant caterpillar that invades Tokyo while searching for the Alilenas, a set of very tiny, twin princesses who have been kidnapped by an evil nightclub owner in the pursuit of big profits. After tiring of crushing buildings and wreaking havoc, the enormous crawly thing zips up into a cocoon and emerges as Mothra, a moth distinguished by both its size and bad attitude. Mothra and the wee princesses make appearances in later Godzilla epics.

Big Bird?

Almost all who investigated the Mothman sightings believed that it was no hoax. The most popular "ordinary" explanation came from West Virginia University biologist Robert Smith, who suggested that the witnesses had seen sandhill cranes. Such cranes are not native to Ohio or West Virginia, but some *could* have migrated down from the plains of Canada.

On November 26, 1966, a small group of people near Lowell, Ohio (70 miles north of Point Pleasant), did report seeing a number of oversized birds in some trees. When approached, the birds flew away and settled on a nearby ridge. From the descriptions—four or five feet tall, with long necks, six-inch bills, and a "reddish cast" in the head area—they *were* probably sandhill cranes. Still, they did not seem to resemble the creature that Mothman witnesses described! In fact, all who saw the monster rejected the sandhill crane identification.

On the other hand, Keel suspected that in a small number of Mothman sightings, excitable observers—frightened by the stories that they had heard—might have mistaken owls seen briefly on dark country roads for something more extraordinary. Regardless, Mothman still resisted easy explanations, for unlike many other monsters, this one had a lot of evidence behind it: a great number of multiple-witness sightings by people that investigators and police officers considered very reliable.

Mothman Across the Sea

Mothman's one known appearance outside Ohio and West Virginia was in England, along a country road near Sandling Park, Hythe, Kent,

John Alva Keel
_____

(1930-)

John Alva Keel is one of the most widely read, influential, and controversial writers on mysterious happenings. Like many anomalists, as a young man Keel was influenced by Charles Fort, the often outrageous theorist and collector of anomalies (also see entry: Falls from the Sky).

Along with many other writings, Keel produced two major books in the early 1970s, _UFOs: Operation Trojan Horse_ and _The Mothman Prophecies_. Though these works deal with UFOs, Keel denies that he is a ufologist. His ideas are, in fact, closer to occultism (the study of supernatural powers) than science, and many consider him a demonologist (one who studies evil spirits).

Keel theorizes that ultraterrestrial gods (gods from another reality beyond our knowledge) once lived on, and ruled, the earth. They left when an early form of the human species began to populate the planet, but, unhappy to have to leave, the gods warred against early humans. Later, some of the gods tried to enlist Neanderthals in their war, and interbreeding between the ultraterrestrials and Neanderthals resulted in the human race as we know it.

Keel believes that humanity's long history of interaction with the supernatural proves the existence of the gods of old as well as the modern Judeo-Christian God. To Keel men in black, monsters, UFOs, and even Mothman are modern versions of the devils and demons of past times.

on November 16, 1963. Four young people reportedly saw a "star" cross the night sky and disappear behind trees not far from them. Frightened, they started to run but stopped soon afterward to watch a golden, oval-shaped light floating a few feet above a field about 80 yards from them. Then the UFO moved into a wooded area and was lost from view.

Suddenly, the observers saw a dark shape shuffling toward them from across the field. It was black, human-sized, and head-

Sandhill cranes—although they fit the description, eyewitnesses saw no resemblance to Mothman.

less, and it had wings that looked like a bat's. At this point the four witnesses left running!

Other people sighted a similar UFO over the next few nights. On November 23, two men who had come to take a look at the area found a "vast expanse of bracken [large coarse ferns] that had been flattened." They also claimed to have seen three huge footprints, two feet long and nine inches wide, pressed an inch deep into the soil.

Sources:

Barker, Gray, *The Silver Bridge,* Clarksburg, West Virginia: Saucerian Books, 1970.
Keel, John A., *The Mothman Prophecies,* New York: E. P. Dutton and Company, 1975.
Keel, John A., *Strange Creatures from Time and Space,* Greenwich, Connecticut: Fawcett Gold Medal, 1970.

BLACK DOGS

A complex, worldwide folklore surrounds certain black dogs, supernatural canines that frequently appear at crossroads and are sometimes connected to the underworld. The lore is most fully documented in Britain, although the American South has its own colorful oral tradition among African Americans. In Mississippi in the early part of the twentieth century, black countryfolk told folklorist N. N. Puckett about huge black dogs with "big red eyes glowing like chunks of fire."

Most tales and reports of supernatural canines describe them as black, but white, gray, and yellow dogs also figure in some stories. Typically, a black dog meets a traveler on a dark road and either guides him to safety or threatens him. Or its appearance may be a sign of the witness's coming death. Black dogs may also attach themselves to families; this idea inspired Sir Arthur Conan Doyle to write the most famous of his Sherlock Holmes novels, *The Hound of the Baskervilles*. Black dogs are said to have glowing eyes, and they often vanish in an instant. Sometimes, especially in accounts from long ago, the black dog is a shape-shifter, at some point revealing his true identity as the devil.

> Sometimes, especially in accounts from long ago, the black dog is a shape-shifter, at some point revealing his true identity as the devil.

Sightings

In his writings, Theo Brown, a leading expert on black-dog lore, has suggested that actual events may be behind a number of legendary tales; indeed, black-dog sightings by reliable witnesses are plentiful.

The first recorded black-dog sighting can be found in a French manuscript, *Annales Franorum Regnum,* dating from 856. The author reported that after a sudden darkness fell over a local church midway through a service, a large dog with fiercely glowing eyes appeared. It dashed about as if searching for something, then suddenly vanished. On August 4, 1577, in Bongay, England, a black dog entered a church during a violent storm, ran through the aisle, killed two worshippers, and injured another by burning him severely. That same day a similar attack occurred at a church in Blibery, seven miles away. These events were related by Abraham Fleming—who had been an eyewitness to the Bongay rampage—and published soon after in *A Straunge Wunder in Bongay.*

Twentieth-century reports of black dogs tend to be less dramatic. Many seem simply to be a variety of ghost story. Typical of these is the account Theodore Ebert of Pottsville, Pennsylvania, gave folklorist George Korson in the 1950s:

REEL LIFE

The Hound of the Baskervilles, 1939.

The curse of a demonic hound threatens descendants of an English noble family until Sherlock Holmes and his faithful assistant Dr. Watson solve the mystery.

One night when I was a boy walking with friends along Seven Stars Road, a big black dog appeared from nowhere and came between me and one of my pals. And I went to pet the dog, but it disappeared from right under me. Just like the snap of a finger it disappeared.

In fact there is no shortage of modern-day black-dog sightings, at least from the early decades of the twentieth century. Englishwoman Ethel H. Rudkin collected a number of reports from her native Lincolnshire and published them in 1938 in *Folklore*. She felt that having had a black-dog experience of her own helped her "get such good first-hand stories." She also wrote, "I have never yet had a Black Dog story from anyone who was weak in body or mind." Unlike ghostly canines elsewhere, those described by Rudkin seemed to have gentle natures.

More recently, what appeared to be a Great Dane reportedly stepped in front of a moving car on Exeter Road in Okehampton, England, on October 25, 1969. Before the driver could stop, the car passed through the animal, which then disappeared! In April 1972 a member of Britain's coast guard saw a "large, black hound-type dog on the beach" at Great Yarmouth. "It was about a quarter of a mile from me," he told the *London Evening News*. "What made me watch it was that it was running, then stopping, as if looking for someone. As I watched, it vanished before my eyes."

Witnesses often mention the creature's glowing eyes. Occasionally, they see little more than the eyes but guess for one reason or another that they belong to a ghostly dog. In the early 1920s young Delmer Clark of La Crosse, Wisconsin, saw "something that looked with shining eyes, with the face of a dog"; in the darkness he thought he could make out a "dark black body." When he saw the creature again a week later in the same location near his home, he kicked at it, only to find his foot inside its mouth! The creature seemed to be expecting it. When Clark screamed, the "dog" disappeared. "I can still see it clearly as I talk now," the man remarked, recalling the incident for his son, author Jerome Clark, in 1976.

In a small number of reports, black dogs are linked—directly or indirectly—to UFOs. One such case was reported in South Africa in 1963. Two men driving at night on the Potchefstroom/Vereeniging road observed a large, doglike animal crossing the highway in front of them. Moments later a UFO appeared and buzzed their vehicle several times, sending them on a frantic escape. Most likely there *was* a large dog in the vicinity, and only coincidence tied it to the UFO. In a case a decade later, though, several Georgia youths claimed that they saw "10 big, black hairy dogs" run from a landed UFO and through a cemetery in Savannah.

Sources:

Bord, Janet, and Colin Bord, *Alien Animals,* Harrisburg, Pennsylvania: Stackpole Books, 1981.

Clark, Jerome, and Loren Coleman, *Creatures of the Outer Edge,* New York: Warner Books, 1978.

BLUES SINGER'S PACT WITH A HELLHOUND

In the 1930s Mississippi native Robert Johnson, the great folk-blues singer/guitarist, did not deny rumors that he had gotten his talents in a midnight deal with a man in black (the devil) whom he met at a crossroads. Johnson even hinted at this event in his 1936 recording "Cross Road Blues." He further commented on this pact in another blues song the following year: "I've got to keep movin'.... There's a hellhound on my trail."

CRAWFORDSVILLE MONSTER

According to a story in the September 5, 1891, issue of the *Indianapolis Journal,* at 2 o'clock on the previous morning in Crawfordsville, Indiana, a "horrible apparition" appeared in the western sky, seen by two men hitching up a wagon. One hundred feet in the air, 20 feet long, and eight feet wide, the headless, oblong thing moved itself along with several pairs of fins and circled a nearby house. It disappeared to the east for a short time and then returned. As curious as the two men were about the strange creature, they decided to run in the opposite direction! They were not, however, the only witnesses. A Methodist pastor, G. W. Switzer, and his wife also observed the monster.

The creature returned the following evening, and this time hundreds of Crawfordsville citizens saw its wildly flapping fins and flaming red "eye." The monster "squirmed as if in agony" and made a

At one point it swooped over a band of onlookers, who swore they felt its "hot breath."

"wheezing, plaintive sound" as it hovered at 300 feet. At one point it swooped over a band of onlookers, who swore they felt its "hot breath."

Many years later anomalist (collector and cataloger of reports of strange physical events) Charles Fort came across the story in a September issue of the *Brooklyn Eagle.* He doubted the account and was "convinced that there had probably never been a Reverend G. W. Switzer of Crawfordsville." Still, he investigated and to his surprise found that the reverend did exist. He wrote to the man, who promised to send Fort a full description of his sighting as soon as he got back from some travels. Unfortunately, Fort reported, "I have been unable to get him to send that account.... The problem is: Did a 'headless monster' appear in Crawfordsville, in September, 1891?"

In time Vincent Gaddis, a Crawfordsville newspaper reporter and writer about unexplained events, would be able to find out more. He interviewed the town's older residents, who confirmed that the story was true and told him about the September 6 group sighting, which had not been reported in the press. Gaddis wrote that "all the reports refer to this object as a living thing"—in other words, it resembled one of the atmospheric life forms, or "space animals" (see box).

Sources:

Fort, Charles, *The Books of Charles Fort,* New York: Henry Holt and Company, 1941.

ONZA

The onza is Mexico's most famous mystery cat, reported for centuries in the remote Sierra Madre Occidental in the northwestern part of the country. Though its existence has not been *officially* recognized by zoologists, few doubt that the animal is, indeed, real.

A Wolflike Cat

To the Aztecs, the onza—or *cuitlamiztli,* as they called it—was a separate animal from the two other large cats, the puma and the jaguar, that lived in the region. After the Spanish conquerors arrived, they called on Aztec emperor Montezuma, who showed them his great zoo. Spaniard Bernal Diaz del Castillo noted that besides jaguars and pumas, he observed another type of big cat that "resembled the wolf."

The later Spanish settlers of northwestern Mexico noted the presence in the wild of a wolflike cat—with long ears, a long, narrow body, and long, thin legs. They gave it the name *onza,* referring to the cheetahs of Asia and Africa that it resembled. They also remarked on its fierceness. "It is not as timid as the [puma]," Father Ignaz Pfefferkorn, a Jesuit missionary stationed in Sonora, wrote in 1757, "and he who ventures to attack it must be well on his guard." According to Father Johann Jakob Baegert, who worked with the Guaricura Indians in Baja California in the mid-eighteenth century, "One onza dared to invade my neighbor's mission while I was visiting, and attacked a 14-year-old boy in broad daylight and practically in full view of all the people; and a few years ago another killed the strongest and most respected soldier" in the area.

Yet outside northwestern Mexico, the onza was practically unknown. The few mentions of it in print attracted no attention, and zoologists continued to believe that only pumas and jaguars lived in the area. No serious scientific expeditions into the rugged country—which in many places was even too wild to reach on horseback—were ever undertaken to try to answer the onza question.

Then, in the 1930s, two experienced hunting guides, Dale and Clell Lee, were working in the mountains of Sonora when

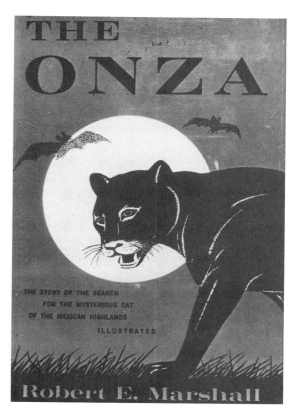

Front cover of Robert Marshall's book *The Onza.*

JAGUAR

A *jaguar* is a large cat (*Felis onca*) of tropical America that is bigger and stockier than the Asian/African leopard and is tan or beige with black spots.

Although there are similarities, the onza's features differ from those of the puma, pictured here.

PUMA

A puma (also called a cougar, mountain lion, panther, or catamount) is a large powerful tan cat (scientific name, *Felis concolor*) that was once widespread in North and South America but is now extinct in many areas.

they heard about the onza for the first time. Some time later, while taking Indiana banker Joseph H. Shirk to hunt jaguars on the wildlife-rich La Silla Mountain, they treed and killed a strange cat that they immediately realized was something they had never seen before. In fact, it looked exactly like the onzas that local residents said lived in the region. After measuring and photographing the animal, they butchered it. Shirk kept the skull and skin. Their present whereabouts are unknown.

Sure that they had found something important, the Lees described the animal to American zoologists. They were stunned when both the scientists and newspaper

Other Fantastic Creatures

accounts laughed at their story. Not accustomed to having their honesty questioned, the upset brothers refused to discuss the experience ever again—that is, until the 1950s, when an Arizona man named Robert Marshall became friends with Dale Lee and thoughtfully recorded the man's testimony. Marshall even went down to Mexico to investigate further. He wrote about his trip in the 1961 book *The Onza*. Except for a single review in a scientific journal, the work attracted no attention at all.

A Modern Look at the Onza

In 1982, at a meeting at the Smithsonian Institution in Washington, D.C., the International Society of Cryptozoology (ISC) was formed. Now, for the first time, biological scientists interested in unknown, unrecognized, or doubted animals had a formal organization through which research could be done. Cryptozoologists, in fact, were among the very few people outside of northwestern Mexico who had heard about the onza.

Ecologist and ISC secretary J. Richard Greenwell lived in Tucson, Arizona. When he learned that Dale Lee and Robert Marshall also lived there, his interest in the onza grew. Marshall showed Greenwell an onza skull that he owned and gave him a cast of its row of upper teeth. Greenwell then took the cast to a West German mammal expert, Helmut Hemmer, who suspected that onzas were leftovers of a prehistoric species of North American cheetah, the *Acinonyx trumani*. While comparisons with fossil skulls eventually ruled out this identification, the case of the mysterious onza was finally beginning to stir scientific interest.

While searching without success for the long-missing onza skull given to Shirk back in the 1930s, Greenwell and Marshall joined forces with two mammalogists (scientists who study mammals) who were also interested in the subject: Troy Best of the University of New Mexico and E. Lendell Cockrum of the University of Arizona. Through Cockrum they met a Mexican rancher who had a perfect skull of an onza killed by another rancher, Jesus Vega. Meanwhile Best, an expert on pumas, had located another onza skull in the Academy of Natural Sciences in Philadelphia.

Then, at 10:30 on the evening of January 1, 1986, two deer hunters in Sinaloa, Mexico, shot and killed a large cat. It was clearly not a jaguar, and they had no idea what it was. Recalling that a few months earlier a rancher friend had talked with visiting scientists about an unusual animal, they alerted Manuel Vega, who recognized the crea-

Recalling that a few months earlier a rancher friend had talked with visiting scientists about an unusual animal, they alerted Manuel Vega, who recognized the creature as an onza as soon as he saw it.

ture as an onza as soon as he saw it; indeed, it was Vega's father who had shot the onza whose skull had brought the scientists there in the first place.

Through the help of a wealthy local family, the body of the animal was placed in a freezer at a large fish company in Mazatlan, and Greenwell was called. Greenwell and Best arrived and, over time, photographed and dissected the creature at a regional government laboratory located in the city. Greenwell wrote: "Upon inspection, the cat, a female, appeared to be as described by the native people. It had a remarkably gracile [slim and graceful] body, with long, slender legs and a long tail. The ears also seemed very long for a puma ... and small horizontal stripes were found on the inside of the forelimbs, which as far as has been determined to date, are not found in puma." Greenwell added that the animal appeared to be about four years old and weighed below the range for adult female pumas. Its total length, however, was normal for a female puma—except for the unusually long tail.

Tissue samples and organs were taken to the United States for further study. But in the following years, Greenwell and Best were too busy with other projects to give much time to the matter. A quick comparison of the onza's tissue samples to those of pumas was done at Texas Tech University; it revealed many similarities and no major differences. Still, conclusions cannot be drawn from a single test of this sort, since animals of different species are often closely related genetically.

Consistent scientific indifference to the onza is puzzling. According to one account, in 1986 ranchers trapped an onza in northern Sonora, near the Arizona border, and kept it alive for a few days while they tried to get officials to take a look at it. Finally, after none expressed interest, its keepers killed the animal and disposed of its body in the dry bed of a stream.

Sources:

Marshall, Robert, *The Onza,* New York: Exposition Press, 1961.

RI

While conducting field studies in New Ireland, an island of Papua New Guinea, University of Virginia cultural anthropologist Roy Wagner heard stories of strange water creatures. Called "ri" (pronounced *ree*) by the local residents, the animals were believed to be air-breathing mammals; they appeared regularly off the island's central and southern coasts. Natives described them as quite human in appearance—except that their torsos had no legs and ended in a pair of side fins; they compared ri to the mermaids on tuna fish cans.

In November 1979, from the coastal village of Ramat, Wagner saw what a native told him was a ri several hundred yards out in Pamat Bay. Wagner recorded the sighting: "Something large [was] swimming at the surface in a broad arc toward the shore. We watched as it came closer, and the best view I got was of a long, dark body swimming at the surface horizontally. Suddenly, a sawfish jumped immediately in front of it, ... and the dark object submerged and did not reappear."

Wagner interviewed a number of islanders who said that they had eaten ri flesh. They did not consider the creatures to be intelligent beings like humans. The ri communicated by whistling and fed on fish. Wagner was certain that his informants were not confusing the animals with dugongs, the plant-eating water mammals also called "sea cows." He was also sure that the islanders were not confusing ri with dolphins.

Wagner's report, published in the first issue of *Cryptozoology* (the journal of the International Society of Cryptozoology or ISC), created quite a sensation. In the summer of 1983 Wagner, ISC secretary J. Richard Greenwell, and two other men traveled to New Ireland, interviewed witnesses, and saw a ri themselves. The sighting took place on the afternoon of July 5, from the village of Nokon on Elizabeth Bay. Every ten minutes the creature, who was plainly feeding, would surface for a few seconds. Because its appearances were brief, decent photographs of the animal proved impossible. It looked to be five to seven feet long, skinny, and had a mammal's tail. Attempts to capture a specimen using a net met with difficulty and had to be given up. The expedition members had other, though briefer, sightings of what appeared to be the same animal.

Ri vs. the Dugong

The researchers returned to the United States convinced that the ri was some kind of unusual animal, not a known animal to which the

Natives described them as quite human in appearance—except that their torsos had no legs and ended in a pair of side fins; they compared ri to the mermaids on tuna fish cans.

A dugong.

local people had attached fantastic features or abilities. Still, they noted that villagers farther north did regard the dugong and the ri as the same creature. Other islanders, however, insisted that they were different—and the investigators tended to agree. The ri stayed underwater as long as ten minutes, while all written information about the dugong had it coming up for air every one or two minutes. Marine biologist Paul Anderson, a specialist in dugongs, sent Greenwell a film of these animals surfacing. Greenwell thought that the shape and actions of the dugongs did not resemble those of the creatures he and his expedition partners had seen.

Mystery Solved

A second expedition in February 1985 solved the mystery. A well-equipped group sponsored by the Ecosophical Research Association saw a ri in Nokon Bay from the deck of the diving ship *Reef Explorer.* When the animal dived, Captain Kerry Piesch (and ship) followed it underwater and took three photographs. Slightly over five feet long, the animal was greenish-gray in color, with a neckless head and short, paddle-like limbs. It was indeed a dugong!

Expedition member Tom Williams later wondered how stories of the ri—another chapter in the age-old myth of merfolk (mermaids and mermen)—could have started and continued "in the face of the obvious reality of the dugong." He recognized, however, that something in human nature and imagination leaps beyond the clear facts and reasonable explanations that our senses and science provide. After this second expedition, Greenwell concluded, "Although we have not found a new species, we have uncovered new data on dugong behavior in deeper water."

Sources:

Wagner, Roy, J. Richard Greenwell, Gale J. Raymond, and Kurt Von Nieda, "Further Investigations into the Biological and Cultural Affinities of the Ri," *Cryptozoology* 2, winter 1983, pp. 113-125.

Williams, Thomas R., "Identification of the Ri Through Further Fieldwork in New Ireland, Papua New Guinea," *Cryptozoology* 4, 1985, pp. 61-68.

FURTHER INVESTIGATIONS

BOOKS

Alien Contacts

Adamski, George, *Inside the Space Ships*, New York: Abelard-Schuman, 1955.

In 1952 Adamski, a UFO writer and photographer of space ships, reported that he met a flying saucer pilot from Venus. He then embarked on a colorful career as a contactee with connections on Mars, Venus, and Saturn. In 1954 a Venusian scoutcraft allegedly flew Adamski around the moon, and this book details his lunar odyssey.

Hopkins, Budd, *Intruders: The Incredible Visitations at Copley Woods*, New York: Random House, 1987.

Hopkins is best known for his UFO-abduction reports, which he, more than any other writer or investigator, has brought to wide public attention. *Intruders,* like his 1981 *Missing Time,* recounts the stories (many evoked under hynosis) of witnesses who were abducted by large-headed, gray-skinned humanoids.

Strieber, Whitley, *Communion: A True Story*, New York: Beach Tree/William Morrow, 1987.

The best-selling UFO book of all time recounts the author's experiences with "visitors"—small, almond-eyed, gray-skinned humanoid occupants of UFOs. A fairly well-known writer of Gothic and futuristic fiction, Strieber contacted UFO-abduction investigator Budd Hopkins after a strange but barely remembered alien contact experience. He wrote this book after hypnosis, which revealed several visitor-related events in his life. William Morrow paid

Strieber $1 million for the book, which attracted enormous attention and a huge reading audience. The film version, starring Christopher Walken as Strieber, met with a fairly tepid response.

Ancient Astronauts

Temple, Robert K. G., *The Sirius Mystery,* New York: St. Martin's Press, 1977.

If the ancient astronaut fad of the 1970s produced one book of substance, many agree this is the one. Learned and extensively researched, it presents a complex, many-sided argument for an early extraterrestrial presence in West Africa.

Von Däniken, Erich Anton, *Chariots of the Gods?: Unsolved Mysteries of the Past,* New York: G. P. Putnam's Sons, 1970.

Along with several other writers in the 1960s, Swiss writer von Däniken theorized in this best-selling book that the gods of Judaism, Christianity, and other religions were extraterrestrials who, through direct interbreeding with our primitive ancestors or through direct manipulation of the genetic code, created *Homo sapiens.* These beings were also responsible for the archaeological and engineering wonders of the ancient world, as well as the mysterious Nazca lines. By no means the original text on the ancient astronaut theory, von Däniken's book took the world by storm, spawning a multitude of sequels, other books, and a popular film on the topic.

Anomalies, general

Fort, Charles, *The Books of Charles Fort,* New York: Henry Holt and Company, 1941.

Until Fort, the pioneer of unexplained physical phenomena, began his extensive research into anomalies, no one knew how "ordinary" and frequent strange happenings really were. This 1941 collection contains *Book of the Damned* (1919), a highly celebrated book that first exposed the reading public to giant hailstones, red and black rains, falls from the sky, unidentified flying objects, and other anomalies. Also in the volume are *New Lands* (1923), *Lo!* (1931), and *Wild Talents* (1932). Along with collecting and recording anomalies, these books present Fort's famously outlandish "theories," which he satirically regarded as no less preposterous than those scientists were offering to explain anomalies.

For further investigation of general anomalies see:

Bord, Janet, and Colin Bord, *Unexplained Mysteries of the 20th Century,* Chicago: Contemporary Books, 1989.
Cohen, Daniel, *The Encyclopedia of the Strange,* New York: Dorset Press, 1985.
Coleman, Loren, *Curious Encounters: Phantom Trains, Spooky Spots and Other Mysterious Wonders,* Boston: Faber and Faber, 1985.

Corliss, William R., ed., *Handbook of Unusual Natural Phenomena,* Glen Arm, MD: The Sourcebook Project, 1977.

Knight, Damon, *Charles Fort: Prophet of the Unexplained,* Garden City, NY: Doubleday and Company, 1970.

Michell, John, and Robert J. M. Rickard, *Living Wonders: Mysteries and Curiosities of the Animal World,* London: Thames and Hudson, 1982.

Bermuda Triangle

Berlitz, Charles, with J. Manson Valentine, *The Bermuda Triangle,* Garden City, NY: Doubleday and Company, 1974.

The Bermuda Triangle fever peaked with the publication of this best-selling book, which sold over five million copies worldwide. Like most of the Triangle books, there is little evidence of original research in its account of the disappearances of planes and boats off the Florida coast, and many of the "facts" that created the mystery were later discredited.

Kusche, Lawrence David, *The Bermuda Triangle Mystery—Solved,* New York: Harper & Row, 1975.

A thorough debunking of what Kusche calls the "manufactured mystery" of the Bermuda Triangle disappearances. For this book, Kusche did the research other Triangle writers had neglected. Weather records, newspaper accounts, official investigators' reports, and other documents indicated that previous Triangle writers had played fast and loose with the evidence.

Cattle Mutilations

Kagan, Daniel, and Ian Summers, *Mute Evidence,* New York: Bantam Books, 1984.

The authors traveled extensively through the western United States and Canada researching the bizarre stories of cattle mutilations. This definitive account exposes journalistic sensationalism and mass hysteria as the only solid basis for the modern-day myth.

Crop Circles

Delgado, Pat, and Colin Andrews, *Circular Evidence,* London: Bloomsbury, 1989.

Delgado, Pat, and Colin Andrews, *Crop Circles: The Latest Evidence,* London: Bloomsbury, 1990.

Two best-selling books on the English phenomenon. The authors, who are active crop circle investigators and founders of the Circles Phenomenon Research group, do not believe that the scientific explanations offered for the mystery so far—either weather- or hoax-related—can begin to explain the anomaly.

Cryptozoology

Bord, Janet, and Colin Bord, *Alien Animals,* Harrisburg, PA: Stackpole Books, 1981.

The best of the Bords' books on paranormal (supernatural) cryptozoology, *Alien Animals* expresses the authors' theory that all mysterious animal sightings, along with UFO sightings and other unexplainable apparitions, are manifestations of "a single phenomenon."

Heuvelmans, Bernard, *On the Track of Unknown Animals,* New York: Hill and Wang, 1958.

Heuvelmans, who is considered the father of cryptozoology, collected innumerable printed references to mysterious, unknown animals from scientific, travel, and popular literature and put them together in this large, informative book that sold over a million copies around the world.

Michell, John, and Robert J. M. Rickard, *Living Wonders: Mysteries and Curiosities of the Animal World,* New York: Thames and Hudson, 1982.

In this lively and literate book, Rickard, founder of the *Fortean Times,* and Michell capture Charles Fort's rich humor and sense of cosmic comedy, while providing encyclopedic, world-ranging coverage of current and historic anomalies.

For further investigation of cryptozoology see:

Caras, Roger A., *Dangerous to Man: The Definitive Story of Wildlife's Reputed Dangers,* New York: Holt, Rinehart and Winston, 1975.

Clark, Jerome, and Loren Coleman, *Creatures of the Outer Edge,* New York: Warner Books, 1978.

Mackal, Roy P., *Searching for Hidden Animals: An Inquiry into Zoological Mysteries,* Garden City, NY: Doubleday and Company, 1980.

Skuker, Karl P. N., *Mystery Cats of the World: From Blue Tigers to Exmoor Beasts,* London; Robert Hale, 1989.

South, Malcolm, ed., *Mythical and Fabulous Creatures: A Source Book and Research Guide,* New York: Greenwood Press, 1987.

Extinct Animal Sightings

Mackal, Roy P., *A Living Dinosaur?: In Search of Mokele Mbembe,* New York: E. J. Brill, 1987.

After two expeditions to the Congo to investigate reports of the mysterious mokele mbembe, University of Chicago biologist Roy Mackal persuasively argues in this book that the monster so frequently reported in this remote area of Africa is in fact some form of sauropod, a dinosaur thought to have been extinct for millions of years.

For further investigation of extinct animal sightings see:

Doyle, Sir Arthur Conan, *The Lost World* (fiction), London: Hodder and Stoughton, 1912.

Guiler, Eric R., *Thylacine: The Tragedy of the Tasmanian Tiger,* Oxford, England: Oxford University Press, 1985.

Folklore

Benwell, Gwen, and Arthur Waugh, *Sea Enchantress: The Tale of the Mermaid and Her Kin,* New York: The Citadel Press, 1965.

This highly regarded book on merfolk lore examines the many traditions of merfolk sightings throughout history and concludes that merfolk must be some kind of unknown, unrecorded species of sea animal.

Evans-Wentz, W. Y., *The Fairy-Faith in Celtic Countries,* New York: University Books, 1966.

The author, an anthropologist of religion, traveled throughout the British Isles recording oral traditions of fairy belief. The resulting book, a classic in folklore studies, also presents the author's extensive research on the existence of "such invisible intelligences as gods, genii, daemons, all kinds of true fairies, and disembodied men."

For further investigation of folklore see:

Otten, Charlotte F., ed. *A Lycanthropy Reader: Werewolves in Western Culture,* New York: Dorset Press, 1986.

Government Cover-ups

Moore, William L., with Charles Berlitz, *The Philadelphia Experiment: Project Invisibility—An Account of a Search for a Secret Navy Wartime Project That May Have Succeeded—Too Well,* New York: Grosset and Dunlap, 1979.

A popular book about the bizarre Philadelphia Experiment during World War II when, according to the letters of a very questionable character, a ship was made invisible and instantaneously transported between two docks, causing its crew members to become insane. The unflagging interest of three Office of Naval Research officers with the far-fetched story was the most unexplainable mystery here. Moore's book inspired the 1984 science fiction movie.

Randle, Kevin D., and Donald R. Schmitt, *UFO Crash at Roswell,* New York: Avon Books, 1991.

The government cover-up of the 1947 "Roswell incident" was effective enough to submerge the story of the crashed flying saucer for nearly 30 years, but the investigation is not closed. Randle and Schmitt have joined others in collecting the testimony of hundreds of witnesses, from local ranchers to air force generals. From these reports they have reconstructed a complex series of events, and the Roswell incident has become one of the best-documented cases in UFO history.

Hairy Bipeds

Napier, John, *Bigfoot: The Yeti and Sasquatch in Myth and Reality,* New York: E. P. Dutton and Company, 1973.

A scientist's view of the abundant evidence of Bigfoot's existence. Napier, a primatologist and the curator of the primate collections at the Smithsonian Institution, was one of the very few conventional scientists to pay serious attention to Bigfoot and other hairy bipeds.

Sanderson, Ivan T., *Abominable Snowmen: Legend Come to Life,* **Philadelphia, PA: Chilton Book Company, 1961.**

The first book to discuss Bigfoot/Sasquatch in any comprehensive manner. Sanderson linked the North American sightings with worldwide reports of "wild men," Almas, and yeti. An encyclopedic view of hairy biped traditions.

For further investigation of hairy bipeds see:

Bord, Janet, and Colin Bord, *The Bigfoot Casebook,* Harrisburg, PA: Stackpole Books, 1982.
Byrne, Peter, *The Search for Big Foot: Monster, Myth or Man?*, Washington, DC: Acropolis Books, 1976.

Lake Monsters

Zarzynski, Joseph W., *Champ: Beyond the Legend,* **Port Henry, NY: Bannister Publications, 1984.**

Vermont's own version of Nessie has a strong advocate in Zarzynski, who formed the Lake Champlain Phenomena Investigation in the 1970s for extensive research of historical sightings and surveillance of the lake. The author links Champ with the Loch Ness monster in this authoritative, if speculative, book.

Loch Ness Monsters

Dinsdale, Tim, *Loch Ness Monster,* **4th edition, Boston: Routledge and Kegan Paul, 1982.**

Over a period of 27 years, British aeronautical engineer Tim Dinsdale made 56 separate expeditions to Ness and spent 580 days watching for the animals. In all, he had three sightings, one of which he filmed. The Dinsdale film is still considered compelling evidence of the existence of the monsters. His highly regarded book, *Loch Ness Monster,* went through four editions between 1961 and 1982.

Holiday, F. W., *The Dragon and the Disc: An Investigation into the Totally Fantastic,* **New York: W. W. Norton and Company, 1973.**

The author, the most radical of the Nessie theorists, originally suggested that the animals in Loch Ness were enormous prehistoric slugs. In this book he changed to an explicitly occult interpretation: Nessies are dragons in the most literal, traditional sense—they are supernatural and evil.

Mackal, Roy P., *The Monsters of Loch Ness,* **Chicago: The Swallow Press, 1976.**

Mackal was the scientific director of the Loch Ness Phenomena Investigation Bureau from 1965 to 1975 and this book, which grew out of his field work, is a cryptozoological classic.

For further investigation of Loch Ness monsters see:

Bauer, Henry H., *The Enigma of Loch Ness: Making Sense of a Mystery*, Urbana, IL: University of Illinois Press, 1986.

Binns, Ronald, *The Loch Ness Mystery Solved,* Buffalo, NY: Prometheus Books, 1984.

Sea Monsters

Heuvelmans, Bernard, *In the Wake of the Sea-Serpents,* New York: Hill and Wang, 1968.

In the most comprehensive volume ever written on the sea serpent, Heuvelmans analyzes 587 sea-serpent reports. He considers 358 of these to be authentic sightings, 49 hoaxes, 52 mistakes, and the rest lack sufficient detail to analyze. The author theorizes that the term "sea serpent" actually covers several unrecognized marine animals, which he classifies in the conclusive chapter.

Sanderson, Ivan T., *Invisible Residents: A Disquisition upon Certain Matters Maritime, and the Possibility of Intelligent Life under the Waters of This Earth,* New York: World Publishing Company, 1970.

Zoologist Sanderson demonstrates his wide-ranging curiosity and his creative imagination in this book in which he theorizes that an intelligent, technologically advanced civilization lives, undetected by the rest of us, in the oceans of the earth. These beings may be extraterrestrials, according to Sanderson, and some UFOs may be their versatile submarines.

For further investigation of sea monsters see:

Lester, Paul, *The Great Sea Serpent Controversy: A Cultural Study*, Birmingham, England: Protean Publications, 1984.

Unidentified Airships

Cohen, Daniel, *The Great Airship Mystery: A UFO of the 1890s,* New York: Dodd, Mead, and Company, 1981.

An entertaining, informative look at the famous turn-of-the-century UFO wave.

UFOs

Hynek, J. Allen, *The UFO Experience: A Scientific Inquiry,* Chicago: Henry Regnery Company, 1972.

At one time a consultant to the air force's UFO-debunking mission, astronomer Hynek changed from a skeptical attitude to a solid belief in the reality of UFOs. In his well-received book, *The UFO Experience,* he blasts the air force's UFO evidence-debunking projects and argues persuasively that science would be greatly furthered by an open-minded study of the subject.

Ruppelt, Edward J., *The Report on Unidentified Flying Objects,* Garden City, NY: Doubleday and Company, 1956.

When Lieutenant Ruppelt, an intelligence officer in the air force, took over the air force investigations of UFOs in the early 1950s, he insisted that investigations be carried out without prior judgments about the reality or unreality of UFOs. By the time he left the project two years later, Ruppelt was largely convinced that space visitors did exist. This memoir of his experiences is considered one of ufology's most important books.

Vallee, Jacques, *Passport to Magonia: From Folklore to Flying Saucers,* Chicago: Henry Regnery Company, 1969.

Vallee is one of the leading theorists on UFOs. *Passport to Magonia* marks his shift from scientific examination of UFO evidence to theories that UFO phenomena have their origins in another reality that is beyond the bounds of scientific analysis. Vallee proposes that inquirers need to immerse themselves in traditional supernatural beliefs—in fairies, gods, and other fabulous beings—in order to begin to understand that aliens and UFOs are only the modern manifestation of ancient beings.

For further investigation of UFOs see:

Blevins, David, *Almanac of UFO Organizations and Publications,* 2nd edition, San Bruno, California: Phaedra Enterprises, 1992.

Clark, Jerome, *The Emergence of a Phenomenon: UFOs from the Beginning through 1959—The UFO Encyclopedia,* Volume 2, Detroit, MI: Omnigraphics, 1992.

Clark, Jerome, *UFOs in the 1980s: The UFO Encyclopedia,* Volume 1, Detroit, MI: Apogee Books, 1990.

Weather Phenomena

Corliss, William R., ed., *Handbook of Unusual Natural Phenomena*, Glen Arm, MD: The Sourcebook Project, 1977.

Corliss, William R., ed., *Strange Phenomena,* two volumes, Glen Arm, MD: The Sourcebook Project, 1974.

Corliss, William R., ed., *Tornados, Dark Days, Anomalous Precipitation, and Related Weather Phenomena: A Catalog of Geophysical Anomalies,* Glen Arm, MD: The Sourcebook Project, 1983.

Corliss, a physicist who systematically catalogued more than 20 volumes' worth of anomalies, applies a conservative, scientific approach to bizarre and seemingly unaccountable events. He was particularly interested in unusual weather, and these volumes of his monumental Sourcebook Project are an invaluable resource in this area.

PERIODICALS

FATE

Llewellyn Worldwide, Ltd.
Box 64383
St. Paul, Minnesota 55164
Monthly

Fate was created in 1948 by Ray Palmer, science fiction editor of *Amazing Stories* and *Fantastic Adventures*, and Curtis Fuller, editor of *Flying*. Floods of flying saucer reports swept the nation following Kenneth Arnold's June 24, 1947, sighting of nine fast-moving discs. Palmer had found in his work that these kinds of "true mystery" stories were wildly popular, and, since no mass-circulation periodical devoted to such matters existed, the two editors decided to fill the void in the market.

Fate covered mysteries relating to ufology, cryptozoology, and archaeology, but its greatest emphasis was on psychic phenomena. In the 1950s Palmer sold his share of the magazine to Fuller, who then edited it with his wife, Mary Margaret Fuller. The magazine was the most successful popular psychic magazine of all time, and achieved a peak circulation of 175,000 during the late 1970s. In 1988 Mary Margaret Fuller was replaced as editor by Jerome Clark, and Phyllis Galde later took over. *Fate*'s 500th issue was published in November 1991.

Strange Magazine

Box 2246
Rockville, Maryland 20852
Semiannual

In the introduction to the first issue of *Strange Magazine,* Mark Chorvinsky, the magician and filmmaker who created the publication, wrote: "We range from wild theoretical speculation to cautious skeptics—including every shade of worldview in between. Some of us are philosophers, others investigators and researchers—surrealistic scientists who catalog the anomalous, the excluded, the exceptional." The focus of *Strange* is on physical rather than psychic anomalies, and it includes topics such as cryptozoology, ufology, archaeological mysteries, falls from the sky, crop circles, and behavioral oddities. Chorvinsky, who has a particular interest in hoaxes, has exposed a number of dubious claims, most notably those associated with English magician and trickster Tony "Doc" Shiels, whose widely reproduced photographs of the Loch Ness monster and a Cornish sea serpent had many fooled. *Strange* reflects its editor's attitude toward anomalous phenomena: open-minded but not credulous.

Published semiannually, each issue of *Strange* is 64 pages long, full of lively graphics and well-written, well-researched articles. It is essential reading for all committed anomalists.

MAJOR ORGANIZATIONS
AND THEIR PUBLICATIONS

Ancient Astronaut Society

1821 St. Johns Avenue
Highland Park, IL 60035
Bimonthly bulletin: *Ancient Skies*

The Ancient Astronaut Society was formed in 1973 and is based on the belief that advanced space beings visited the earth early in human's history and possibly played a part in the development of human intelligence and technology. The organization is directed by attorney Gene M. Phillips in Chicago. European director Erich von Däniken, who wrote many books about the idea—including the wildly popular *Chariots of the Gods?*—operates out of Switzerland. The organization publishes a bimonthly bulletin, *Ancient Skies,* in both English and German. It also meets in a different world city every year and sponsors archaeological expeditions to sites where ancient marvels, viewed as evidence for the group's beliefs, can be seen by members firsthand.

Center for Scientfic Anomalies Research (CSAR)

Box 1052
Ann Arbor, MI 48106
Journal: *Zetetic Scholar*

Formed in 1981, the Center for Scientific Anomalies Research is a "private center which brings together scholars and researchers concerned with furthering responsible scientific inquiry into and evaluation of claims of anomalies and the paranormal." Director Marcello Truzzi and associate director Ron Westrum are sociologists of science at Eastern Michigan University in Ypsilanti. From 1978 to 1987 Truzzi, who had cofounded and then resigned from the Committee for the Scientific Investigation of Claims of the Paranormal (see CSICOP entry below), edited the journal *Zetetic Scholar,* a forum in which believers and nonbelievers could discuss and debate their views on anomalous subjects. CSAR was created as a parent organization for the publication; a number of important physical and biological scientists, psychologists, and philosophers are among its consultants.

Committee for the Scientific Investigation
of Claims of the Paranormal (CSICOP)

Box 703
Buffalo, NY 14226
Quarterly magazine: *Skeptical Inquirer*

The Committee for the Scientific Investigation of Claims of the Paranormal was formed in 1976 by Paul Kurtz, a professor of philosophy at the State University of New York at Buffalo, and science sociologist Marcello Truzzi (see CSAR entry above). Unfortunately, the two founders had different aims for the organization right from the start. Kurtz and his followers were rigid disbelievers in ufology, astrology, and other subjects that existed on the fringes of science; more than just skeptics, they viewed such unusual ideas as threats to reason and civilization. Truzzi felt that at least some claims, especially those made by serious parapsychologists, cryptozoologists, and ufologists, were worth investigating. Truzzi had hoped that the organization would practice fairminded *nonbelief* rather than ridiculing *disbelief.* He resigned a year later.

From the beginning CSICOP attracted many famous scientists. The organization was well funded and by the late 1980s its quarterly magazine, the *Skeptical Inquirer,* would claim a circulation of more than 30,000—the world's second most popular magazine on anomalies and the paranormal (after the psychic digest *Fate*). CSICOP sponsors regular conferences. Through its connected publishing house, Prometheus Books, it releases works expressing the debunker's view of UFOs, the Loch Ness monster, the Bermuda Triangle, and other "antiscientific" matters.

Fund for UFO Research, Inc. (FUFOR)

Box 277
Mount Rainier, MD 20712

The Fund for UFO Research was founded in 1979 and provides grants for scientific research and educational projects on UFO-related subjects. Studies have included investigations of UFO photographs and crash reports. The organization publishes findings from these projects from time to time. Funds are granted by a ten-member board of directors of scientists and scholars.

International Fortean Organization (INFO)

Box 367
Arlington, VA 22210

Quarterly journal: *INFO Journal*

The International Fortean Organization was founded in 1965 by Ronald J. Willis. It is dedicated to the memory and interests of pioneering anomaly collector Charles Fort (1874-1932). The *INFO Journal* is a forum for a wide variety of unexplained physical happenings, both past and present. The organization sponsors a yearly "FortFest" in the Washington, D.C., area, where well-known writers and researchers discuss anomalies.

INFO is a successor to the Fortean Society (1931-1960). Under the direction of writer and advertising man Tiffany Thayer, that organization and its *Fortean Society Magazine* (retitled *Doubt* in 1944) were known for their weird ideas and eccentric writers. The society oversaw the important publication of *The Books of Charles Fort,* a collection of the anomalist's writings, in 1941.

International Society of Cryptozoology (ISC)

Box 43070
Tucson, AZ 85733

Yearly journal: *Cryptozoology*

Quarterly newsletter: *ISC Newsletter*

The International Society of Cryptozoology was formed in early 1982. At its founding meeting at the Museum of Natural History of the Smithsonian Institution in Washington, D.C., the organization defined its purpose as the "scientific inquiry, education, and communication among people interested in animals of unexpected form or size, or unexpected occurrence in time and space." Roy P. Mackal, a University of Chicago biologist, and University of Arizona ecologist J. Richard Greenwell had worked behind the scenes for more than a year to put the organization together. (They, along with cryptozoology pioneer Bernard Heuvelmans, became the ISC's elected officers.)

With its serious, scientific approach to the subject of "unexpected" animals, the organization has been able to attract well-respected zoologists, anthropologists, and others as members. The ISC holds an annual meeting, always at a university or scientific institute. It publishes the quarterly *ISC Newsletter* and the yearly journal *Cryptozoology*. Although mainstream science has still not fully accepted cryptozoology, the ISC has enhanced the respectability of the field.

Intruders Foundation (IF)

Box 30233
New York, NY 10011

Yearly bulletin: *IF: The Bulletin of the Intruders Foundation*

The Intruders Foundation was created by Budd Hopkins, author of two popular books on UFO abductions. The purpose of the organization is to fund research and to offer therapeutic help to the many people who have contacted Hopkins, disturbed by their own claimed abduction experiences. The foundation has an informal national network of mental health professionals who volunteer to counsel these people. *IF: The Bulletin of the Intruders Foundation* appears yearly, and discusses abduction cases, investigations, and other related matters.

J. Allen Hynek Center for UFO Studies (CUFOS)

2457 West Peterson Avenue
Chicago, IL 60659

Journal: *Journal of UFO Studies (JUFOS)*

Newsletter: *International UFO Reporter (IUR)*

The Center for UFO Studies was formed in 1973 by J. Allen Hynek, the head of Northwestern University's astronomy department, and Sherman J. Larsen, a businessman and director of a small UFO group in Chicago. During the 1950s and 1960s Hynek had been the U.S. Air Force's chief scientific consultant on UFOs, until he publicly complained that the military was doing a very poor job of investigating reports. While the prevailing air force attitude toward UFOs was one of disbelief and dismissal, Hynek thought that UFOs were probably more than mistaken identifications and hoaxes. CUFOS was created so that scientists and other trained professionals could deal with UFO research in a serious but open-minded way.

CUFOS is one of two major UFO groups in the United States (the other is the Mutual UFO Network; see MUFON entry below). CUFOS publishes a newsletter, the *International UFO Reporter* (*IUR*), and the *Journal of UFO Studies* (*JUFOS*), and has sponsored UFO investigations. Located in Chicago, its huge collection of research materials is available to people studying UFOs. The organization's official name was expanded after Hynek's death in 1986.

Mutual UFO Network, Inc. (MUFON)

103 Oldtowne Road
Seguin, TX 78155

Monthly magazine: *MUFON UFO Journal*

Walter H. Andrus, Jr., a former officer of Tucson's Aerial Phenomena Research Organization, founded the Midwest UFO Network in 1969, based in Quincy, Illinois. In 1975 his group moved to Seguin, Texas, and became MUFON, the Mutual UFO Network. One of the most successful UFO organizations in the brief history of UFOs, it would claim 4,000 national and international members by 1992. Though open-minded about different explanations for UFOs, MUFON clearly leans towards the extraterrestrial hypothesis. The organization hosts a conference in a different U.S. city every year. MUFON's monthly magazine, *MUFON UFO Journal* (formerly *Skylook*), contains serious UFO studies and is essential reading for ufologists.

Society for the Investigation of the Unexplained (SITU)

Box 265
Little Silver, NJ 07739

Magazine: *Pursuit*

Founded in 1965 by science writer and lecturer Ivan T. Sanderson, the Society for the Investigation of the Unexplained publishes the magazine *Pursuit* about three or four times a year. *Pursuit* reports on anomalies and looks for the explanations or meaning behind them.

Society for Scientific Exploration (SSE)

Box 3818, University Station
Charlottesville, VA 22903

Semiannual journal: *Journal of Scientific Exploration*

Semiannual newsletter: *Explorer*

The Society for Scientific Exploration was formed in 1982. It encourages scientific study of UFOs, unexpected animals, supernatural claims, and other subjects that lie on the edges of science, because "progress towards an agreed under-standing of such topics ... is likely to be achieved only if they are subject to the normal processes of open publication, debate, and criticism which constitute the lifeblood of science and scholarship." Full SSE members must be connected with a major university, government group, or corporate research institution; those who do not qualify are associate members. The society sponsors a conference each year at an American university. It publishes both the newsletter *Explorer* and the *Journal of Scientific Exploration* twice a year.

INDEX

Bold numerals indicate volume numbers.

A

Abductions
    by inner earth people **1:** 57
    by UFOs **1:** 11, 15, 167-168
Abominable snowman **2:** 244-249, 266, 271, 284
Abominable Snowmen: Legend Come to Life **2:** 233, 244, 255
Adams, John **1:** 67
Adamski, George **1:** 12, 45, 47, 63, 167
Advanced civilizations **1:** 30, 58, 64; **3:** 379-380
Aerial Phenomena Research Organization (APRO) **1:** 107, 153
Africa **2:** 295, 302-310
Agnagna, Marcellin **2:** 296, 308-309
Air force **1:** 4-9
Airplanes, mysterious **1:** 154
Airships **1:** 4, 17-21
Air Technical Intelligence Center (ATIC) **1:** 6
Akers, David **1:** 107
Alecton **3:** 388-390
Alien crash victims **1:** 14, 75
Aliens in the clouds **1:** 153
Allen, Carl Meredith. *See* Allende, Carlos Miguel
Allende, Carlos Miguel **1:** 80-83
Alligators
    falls from the sky **1:** 128
    in sewers **2:** 220-222

Almas **2:** 281-284
Among the Himalayas **2:** 265
Ancient astronauts **1:** 25-27, 28, 30, 31-32, 82, 168
Ancient Astronaut Society **1:** 26, 176
Ancient civilizations **1:** 31
Ancient Skies **1:** 27
Andersen, Hans Christian **3:** 465
Andrews, Colin **3:** 505, 506
Animals, extinct **2:** 295-321
Animals, misplaced **2:** 203-228
Animals Nobody Knows **2:** 233
Animal Treasures **2:** 233
Antimatter **1:** 159
Apemen **2:** 231-292
Apes **2:** 231-292
Apollo aliens **1:** 65
Archaeological expeditions **1:** 26
Archaeological sites **1:** 30
Area 51 **1:** 83-84
Arnold, Kenneth **1:** 4, 12, 73, 152
Arnold, Larry E. **3:** 491, 493
Arsenyev, V. K. **1:** 37-38
Astronauts **1:** 66
Astronomy **1:** 67
Astronomy legends **1:** 31
Atlantis **1:** 30, 64
Atmospheric life forms **1:** 133; **2:** 340
Atomic Energy Security Service (AESS) **1:** 111
At the Earth's Core **1:** 57
Avenger torpedo bombers **1:** 89-92

Australia **2:** 289, 291, 315
Aztec Indians **2:** 341

B

Babylonia **2:** 322-324
Baby ... Secret of the Lost Legend **2:** 298
Ballard, Guy Warren **1:** 56
Ball lightning **1:** 155-159
Barisal guns **1:** 160
Barker, Gray **1:** 49
Barnum, P. T. **3:** 428-429
Barry, James Dale **1:** 156
Batman **1:** 37
"Bat-men" **1:** 60
Battelle Memorial Institute **1:** 7
Bauer, Henry H. **3:** 431
The Beast from 20,000 Fathoms **3:** 378
Beast of Exmoor **2:** 215-217
Beast of Gevaudan **2:** 217-220
Beasts and Men **2:** 302
Behemoth **2:** 323
Behind the Flying Saucers **1:** 75
Bender, Albert K. **1:** 48-49
Benjamin, E. H. **1:** 18
Bergier, Jacques **1:** 25, 30
Berlitz, Charles **1:** 66, 96, 169
Bermuda Triangle **1:** 89-97, 98
The Bermuda Triangle **1:** 96
The Bermuda Triangle Mystery—Solved
    1: 96
Bernard, Raymond **1:** 57
Bessor, John Philip **1:** 133; **2:** 340
Bible **2:** 323, 325-326; **3:** 484-485
Bielek, Alfred **1:** 83
Bigfoot **2:** 198-199, 231-232, 241-249, 288
Bigfoot **2:** 240
Bioluminescent organisms **3:** 375
Birdman **1:** 37
Black dogs **1:** 106; **2:** 336
Black panthers **2:** 205
Blavatsky, Helena Petrovna **1:** 26
Bleak House **3:** 490
Blimp sightings **1:** 20
The Blob **1:** 14
Blood and flesh rain **1:** 129, 134-136
"Blue Room" **1:** 75
Book of Genesis **3:** 484
Bord, Colin **3:** 482
Borderland Sciences Research
    Foundation **1:** 58, 81
Bord, Janet **3:** 482
Botetourt gasser **3:** 511
Brontosaurus **2:** 303, 304

"The Brown Mountain Light" **1:** 114
Brown Mountain lights **1:** 114-116

C

Cadborosaurus **3:** 400-401
Caddy **3:** 400-401
Caimans **2:** 222
California Airship Scare **1:** 17-18
The Case for the UFO **1:** 26, 80, 82, 95
Cat People **2:** 213
Cats, strange **2:** 205-214, 215-217,
    340-344
Cattle mutilations **3:** 493-498
CE1s **1:** 10
CE2s **1:** 10
CE3s **1:** 10-11
Center for Crop Circle Studies
    3: 504, 507
Center for Cryptozoology **2:** 197
Center for UFO Studies (CUFOS)
    1: 8-9, 14, 178-179
Cephalopods **3:** 384
Cereology **3:** 498, 505-506
Champ **2:** 198; **3:** 426-434
Champ: Beyond the Legend **3:** 431
Champlain, Samuel de **3:** 427, 429
Chariots of the Gods?: Unsolved Mysteries
    of the Past **1:** 25- 26, 30
Charroux, Robert **1:** 25
Chatelain, Maurice **1:** 66
Chesapeake Bay **3:** 401
Chessie **3:** 401-402
Chiles-Whitted UFO sighting **1:** 5
China **2:** 284
Christian fundamentalists **3:** 484, 486
Christianity **1:** 25
Cigar clouds **1:** 150-152
Circles Effect Research Group (CERES)
    3: 501
Circles Phenomenon Research **3:** 506
Classification of animals **2:** 194
Close Encounters of the Third Kind
    1: 9, 47, 79, 92
Clouds **1:** 149-155
Coast Salish Indians **2:** 241
Coelacanth **2:** 195, 196; **3:** 449
Coins: falls from the sky **1:** 125
Coleman, Loren **2:** 205, 222, 226-227
Collins, George D. **1:** 18
Colored rain **1:** 134
Columbus, Christopher **3:** 466
Comets **1:** 164
The Coming of the Fairies **3:** 461

Condon Committee **1:** 9
Condon, Edward U. **1:** 9
Condors **3:** 478, 479
Congo **2:** 295, 324
Contactees **1:** 12-13, 30, 45-47, 50, 64
Corliss, William R. **1:** 141-142, 149
"Corpse candles" **1:** 104
Cottingley fairy photographs **3:** 459
Cover-ups, government **1:** 9, 14, 63,
    73-85, 171
Crawfordsville monster **2:** 339-340
Creationism **2:** 325
The Creature from the Black Lagoon
    2: 329-330
Creature-haunted lakes **3:** 413
Crocodiles **2:** 221
Cro-Magnon race **2:** 259
Cronin, Edward W., Jr. **2:** 268
Crop circles **3:** 498-508
Cropper, Paul **2:** 212-213
"Cross Road Blues" **2:** 339
Cruz, Carlos Berche **1:** 52
Cryptozoology **2:** 193-200
Cuvier, Georges **2:** 194

D

Dahinden, Rene **2:** 244, 255
Darwin, Charles **2:** 194, 325
The Day the Earth Stood Still **1:** 47
Death candle **1:** 104
Debunking **1:** 5-6, 9
Deep See **1:** 94
Delgado, Pat **3:** 505-507
De Loys, Francois **2:** 273-277
"Demon of the Deep" **3:** 376-377
Destroying rain clouds **1:** 154
Devil's Sea **1:** 97-98
The Devil's Triangle **1:** 96
Dickens, Charles **3:** 490
Didi **2:** 277
Dieterlen, Germaine **1:** 32-33
Dinosaurs **2:** 196, 295-301, 303, 310,
    323-326; **3:** 378
Dinsdale, Tim **3:** 423-425, 520
The Disappearance of Flight 19 **1:** 93
Divining rod **3:** 507
Dogon **1:** 31-33
Dogs, black **2:** 336-339
Dowsing **3:** 505, 507
Doyle, Arthur Conan **2:** 197, 298, 337;
    3: 461-462
Dragons **2:** 303, 322-323; **3:** 410, 412, 449
Drought **1:** 154-155

Dugongs **2:** 198, 345-346; **3:** 426,
    466, 468
Dwarfs **1:** 41-42, 77

E

Edwards, Frank **1:** 96
Egyptian pyramids **1:** 25
Einstein's (Albert) Unified Field Theory
    1: 80
Eisenhower, Dwight D. **1:** 16
Elephant seals **3:** 426, 444
Entombed animals **2:** 222-226
Ethereans **1:** 58-59
Etheric doubles **1:** 58
E.T.: The Extra-Terrestrial **1:** 14
Evans, Hilary **3:** 459, 508
Evans-Wentz, W. Y. **1:** 104; **3:** 454, 519
Evidence of UFOs **1:** 14
Evolution **2:** 194, 325
Exon, Arthur **1:** 79
Exotic animals **2:** 287
The Expanding Case for the UFO **1:** 26, 64
Extinction **2:** 196, 295-326
"Extraterrestrial biological entities"
    (EBEs) **1:** 16
Extraterrestrial communications
    1: 25-33, 45-52
Extraterrestrials **1:** 37-52, 73-79
Extraterrestrials on moon **1:** 61-64

F

Fairies **1:** 13, 104; **3:** 452-459, 503
The Fairy-Faith in Celtic Countries
    3: 454
Fairy Rings **3:** 502, 504
Falling stars **1:** 145
Falling stones **1:** 125
Falls from the sky **1:** 121-133; **2:** 340
The Fire Came By **1:** 163
Fire in sky **1:** 123
Fish: falls from the sky **1:** 125, 127
Flatwoods monster **1:** 42-45
Flesh: falls from the sky. *See* Blood and
    flesh rain
Flight 19 **1:** 89-94
Flying dragons **3:** 447-452
Flying humanoids **1:** 37-40, 77
Flying saucer **1:** 3-16, 58, 85
The Flying Saucer Conspiracy **1:** 94, 96
Flying Saucers and the Three Men **1:** 49
Flying snake **3:** 447, 450-452

Folklore **3:** 410, 447-486
Foo fighters **1:** 4
Fort, Charles **1:** 62-63, 128-130, 132-133, 140-141, 168; **2:** 198, 234, 335, 340; **3:** 415, 486
Fortean Picture Library **3:** 482
Fortean Times **3:** 482
Fortin, Georges **1:** 151
The 4D Man **1:** 59
Fourth dimension **1:** 58-59, 94
Freedom of Information Act **1:** 14
Freeman, Paul **2:** 247-249
Friedman, Stanton T. **1:** 14, 74
Friedrich, Christof. *See* Zundel, Ernst
Frogs
    entombed **2:** 223
    falls from sky **1:** 130
From Outer Space to You **1:** 63

G

Gaddis, Vincent H. **1:** 81, 89, 94, 96, 106; **2:** 340
Galactic Federation **1:** 45
Garcia, Ignacio Cabria **1:** 52
Gardner, Marshall B. **1:** 55-56
GeBauer, Leo A. **1:** 75
Genzlinger, Anna Lykins **1:** 81
George Adamski Foundation **1:** 63
Ghost lights **1:** 103-110
Ghost rockets **1:** 4
Giant octopus **3:** 380-385
Giant panda **2:** 195, 287
Giants **2:** 325
Giant squid **2:** 195; **3:** 382, 388-393
Gimlin, Bob **2:** 245
Globsters **3:** 384-387
Gloucester, Massachusetts **3:** 396-397
Goldwater, Barry **1:** 73
Gorillas **2:** 226, 232, 234, 255
Gould, Rupert T. **3:** 414-416
Government conspiracies **3:** 495
Graham, Francis **1:** 64
Grain: falls from the sky **1:** 130
Gray, Hugh **3:** 419-420
The Great Flood **2:** 325; **3:** 484
The Great Sea-Serpent **2:** 195; **3:** 400
Green fireballs **1:** 111-114
Green, John **2:** 241, 244, 255-257
Green slime: falls from the sky **1:** 131
Greenwell, J. Richard **2:** 198-199, 209, 287-288, 308, 343-347; **3:** 387, 439
The Grey Selkie of Sule Skerrie **3:** 462
Griffiths, R. F. **1:** 139, 140

Group d'Etude des Phenomenes Aero-spatiaux Non-Identifies (GEPA) **1:** 11
Guiler, Eric R. **2:** 317-318

H

Hagenbeck, Carl **2:** 302
Hail **1:** 138-139
Hairy bipeds **2:** 197, 199, 231-292
Hairy dwarfs **1:** 41-42
Hall, Asaph **1:** 69
Halley's comet **1:** 164
Hall, Mark A. **3:** 478
Hallucinations **3:** 459
Hangar 18 **1:** 14, 73-80
Hangar 18 **1:** 75
Harry and the Hendersons **2:** 248
Hart, W. H. H. **1:** 18
Hazelnuts: falls from the sky **1:** 131
Herschel, John **1:** 60
Herschel, William **1:** 60-61
Hessdalen lights **1:** 106-110
Heuvelmans, Bernard **2:** 193, 195-197, 257-261, 272, 302, 304, 324; **3:** 393-394, 402-405, 469
Hillary, Edmund **2:** 268, 270
Himalayan Mountains **2:** 265
Hitler, Adolf **1:** 57
Hoaxes
    airships **1:** 19
    contactees **1:** 46
    crop circles **3:** 505
    hairy bipeds **2:** 240
    Loch Ness monster **3:** 421
    Minnesota iceman **2:** 257-261
    moon **1:** 61-62
    sea serpent **3:** 404
    space brothers **1:** 46-47
    spaceship crashes **1:** 75
    thylacine **2:** 320
    UFO **1:** 12-13
    Ummo **1:** 52
Holiday, F. W. **3:** 424, 520
Hollow earth **1:** 55-58
The Hollow Earth **1:** 57
Holmes, Sherlock **2:** 337
Homo sapiens **1:** 25
Hopkins, Budd **1:** 14, 167
The Hound of the Baskervilles **2:** 337
Hudson, Henry **3:** 467
Human-animal transformations **3:** 470-471
Humanoids **1:** 11, 15, 37-40, 41-42, 153
Hynek, J. Allen **1:** 7-9, 50, 107, 173, 178-179; **3:** 458

I

Ice falls **1:** 137
Indian, American, anomaly reports
    2: 242-243; **3:** 440, 474, 476
Inorganic matter: falls from the sky
    1: 121
Inside the Space Ships **1:** 63
International Flying Saucer Bureau
    (IFSB) **1:** 48
International Fortean Organization
    1: 129, 177
International Society of Cryptozoology
    (ISC) **2:** 193, 197-199, 209, 271, 307,
    309, 343, 360; **3:** 387, 402, 439
International UFO Bureau **2:** 250,
In the Wake of the Sea-Serpents **2:** 197;
    3: 402
Intruders **1:** 15, 167
Intruders Foundation **1:** 15
Invisible Horizons **1:** 81, 94
Invisible Residents **1:** 96; **2:** 233; **3:** 379
In Witchbound Africa **2:** 312
Ishtar Gate **2:** 322, 324
Island of Lost Souls **2:** 219
It Came from Outer Space **1:** 14
Iumma **1:** 50
I Was a Teenage Werewolf **3:** 474

J

Jacko **2:** 254-257
Jaguar **2:** 341
J. Allen Hynek Center for UFO Studies
    (CUFOS) **1:** 14, 178-179
The Jessup Dimension **1:** 81
Jessup, Morris K. **1:** 26, 64, 80-83, 95,
    133, 141
Johnson, Robert **2:** 339
Jones, Mary **1:** 105-106
Journal of American Folklore **1:** 50
Journey to the Center of the Earth **1:** 57
A Journey to the Earth's Interior **1:** 56
Judaism **1:** 25
Jurassic Park **2:** 298

K

Kagan, Daniel **3:** 497-498, 517
Kangaroos, misplaced **2:** 203-205
Kaplan, Joseph **1:** 112-113
Kasantsev, A. **1:** 163
Keel, John A. **1:** 49-50, 59; **2:** 331, 333-335

Kelpies **3:** 410, 414
Keyhoe, Donald E. **1:** 9, 13-14, 64, 94, 96
King, Godfre Ray. *See* Ballard,
    Guy Warren
King Kong **2:** 239, 240, 301
Knight, Damon **1:** 133
Koldeway, Robert **2:** 324
Kongamato **2:** 312
Kraken **3:** 389
Krantz, Grover **2:** 232, 248
Kuban, Glen J. **2:** 325, 326
Kulik, Leonid **1:** 162
Kusche, Larry **1:** 93, 96-97, 98, 169

L

Lake Champlain **3:** 426-427
Lake Champlain Phenomena Investiga-
    tion (LCPI) **3:** 434
Lake monsters **3:** 409-444
Lake Okanagan **3:** 439-441
Lake Tele **2:** 308, 309
Lake Worth monster **2:** 252-254
The Land before Time **2:** 298
La Paz, Lincoln **1:** 111-113
Larsen, Sherman J. **1:** 8
Layne, N. Meade **1:** 58
Lazar, Robert Scott **1:** 85
Leaves: falls from the sky **1:** 131
Lee, Gloria **1:** 13
Legend of Boggy Creek **2:** 248
LeMay, Curtis **1:** 73
Lemuria **1:** 30, 57
Leonard, George H. **1:** 64
Leopard **2:** 206
Lescarbault **1:** 67-68
Leverrier, Urbain **1:** 66-69
Lewis and Clark expedition **1:** 160
Ley, Willy **2:** 196, 304, 322-324
Lights in folk tradition **1:** 103
Limbo of the Lost **1:** 96
Linnaeus, Carolus **2:** 194
Linnean Society **3:** 397-398
The Little Mermaid **3:** 465
"Little people" **1:** 26, 64, 82
A Living Dinosaur? **2:** 307
Living dinosaurs **2:** 197, 295-301
Living Mammals of the World **2:** 233
Lizards: entombed **2:** 225-226
Local lights **1:** 106
Loch Morar **3:** 434-438
Loch Ness **3:** 413-426
Loch Ness and Morar Project
    3: 423, 425

Q

Queensland tiger **2:** 213-214

R

Randle, Kevin **1:** 74
Red wolf **3:** 471
Regusters, Herman **2:** 296, 308-309
Religions **1:** 25
Religious lights **1:** 105-106
The Report on Unidentified Flying Objects
    (1956) **1:** 6
Reptile men **2:** 199, 329-331
Return of the Ape Man **2:** 239
Ri **2:** 198, 345-347
Ridpath, Ian **1:** 32-33
Road in the Sky **1:** 30
The Robertson Panel **1:** 6-7
Robins Air Force Base **1:** 5
Rocky Mountain Conference on UFO
    Investigation **1:** 47
Rogo, D. Scott **1:** 125
Rojcewicz, Peter M. **1:** 50
Roswell incident **1:** 14, 73-80, 171
The Roswell Incident **1:** 66
Rudkin, Ethel H. **2:** 338
Runaway clouds **1:** 150
Ruppelt, Edward J. **1:** 5, 6, 113

S

Sagan, Carl **1:** 32, 33
Salamanders: falls from the sky **1:** 131
Sananda **1:** 13
Sanderson, Ivan T. **1:** 81, 96, 98-99; **2:**
    196-197, 233-234, 244, 255, 257-261,
    267, 297-298, 304; **3:** 379
Sandoz, Mari **3:** 448
Sarbacher, Robert **1:** 78-79
Sasquatch **2:** 241-242
Satan worshippers **3:** 494-495
Saucer clouds **1:** 154
Saucer nests **3:** 503-504, 506
Saucers, flying **1:** 3-16
Sauropods **2:** 295, 302, 310, 323
Scandinavia **3:** 413
Schmidt, Franz Herrmann **2:** 298-300
Schmitt, Don **1:** 75
Science Frontiers **1:** 141
*Scientific Study of Unidentified Flying
    Objects* **1:** 9
Scully, Frank **1:** 75

Sea cows **3:** 426, 468
Sea monsters **3:** 375-380
Searching for Hidden Animals **2:** 307
Sea serpents **2:** 195; **3:** 394
Secret of the Ages **1:** 55, 57
Sedapa. *See* Orang-Pendek
Seeds: falls from the sky **1:** 131
Selkies **3:** 462
Serpents **2:** 322, 323; ; 394-405; 447-452
Sesma, Fernando **1:** 50, 52
Seventh District Air Force Office of
    Special Investigations (AFOSI) **1:** 111
Shape-changing **2:** 337; **3:** 435, 473
Shaver, Richard Sharpe **1:** 57
Shiels, Anthony "Doc" **3:** 404, 421
Shooting stars **1:** 112
Siegmeister, Walter. *See* Bernard,
    Raymond
Sirius **1:** 31
Sirius B **1:** 31
Sirius mystery **1:** 31-33
The Sirius Mystery **1:** 31-33
Sirrush **2:** 322-324
"Sky gods" **1:** 30
Skyquakes **1:** 160-161
Sky serpents **3:** 447-452
Slick, Tom **2:** 268-270, 273
Smith, J. L. B. **3:** 449
Smith, John **3:** 466
Smithsonian Institution **2:** 193, 258-259
Smith, Wilbert B. **1:** 78
Society of Space Visitors **1:** 50
Solar system **1:** 68
Somebody Else Is on the Moon **1:** 64
Sonar **3:** 413, 425, 432
South Pole **1:** 57
Space animals **1:** 133; **2:** 340
Space brothers **1:** 12, 45-47
Space Review **1:** 48
Spencer, John Wallace **1:** 96
Sperm whales **3:** 391, 393
Spielberg, Steven **1:** 9, 79, 92
Splash **3:** 465
Spontaneous generation theory **1:** 121
Spontaneous human combustion
    3: 489-493
Sprinkle, R. Leo **1:** 47
Squids **3:** 388-393
Squid-whale battles **3:** 391-393
Star jelly **1:** 142-145
Starman **1:** 47
Steckling, Fred **1:** 63
St. Elmo's fire **1:** 156
Stewart, Jimmy **2:** 270